The

Time Editors

by

Luke Lang

For Mae,

The one person who will

always want the best for

me no matter what.

I Love You, Mama!

Author's Note

If you're a history buff looking for someone else's take on an alternate timeline, you've got the wrong book. My first (and only) 'D-' in my scholastic career was in 12th grade Advanced History. Actually I had a 57, but I studied really hard and scored a 98 on the final exam. Maybe if I'd studied throughout the year, I might have, eh...

This is my first attempt at writing a book. I stared at the ceiling one night and made up my own story. Before the night was over, this is what came out, almost like word vomit. It even scared me a bit. Enjoy!

The

Time Editors

"We are irrefutably the sum of our experiences. The decisions we make throughout our lives mold and shape us into the people we are today, and the decisions we make today will affect who we are tomorrow."

- Barton Urthorn, Head of the Assembly

Chapter 1

Order in the Court

The Prime Hall filled with frightened Solayans behaving like a classroom of unruly children. The air buzzed with worried voices trying to ascertain the purpose of the gathering. Up in the balconies, to the midlevel seating, and down on the main floor, all available seating had been extended. At the rate Solayans entered, the Prime Hall's maximum capacity rating was soon to be tested.

Between the massive supply of historical data in the archives and his vast knowledge of history, LaDon Grafter could not recall any time the Prime Hall held so much angst at once. Even in his grandfather's stories, nothing compared to this moment. And if anyone on the planet was more suited to have known such a fact, it was LaDon.

Other than his extensive knowledge of history and obsession with the Assembly, LaDon was a well-rounded young Solayan. He kept in shape when he got the chance, although he didn't have to try too hard. He sported a medium build, dark hair, and a handsome face, even though, most of the time, his face was buried in work. At only thirty-three, he was one of Solaya's top historians. Why wouldn't he be? After the death of his beloved parents when he was just a baby, LaDon was raised by his grandfather, Pomph Grafter, one of the most famous historians in the world.

He had finished up at his terminal and hurried over to the Prime Hall as soon as he received the communication. A gathering called by the Assembly with such short notice was sure to be a big deal. The verbiage within the communication pointed toward another scientific finding, but didn't elaborate any further. Usually, the Assembly would release a public memo about the find and life would go on. To call a meeting in the Prime Hall definitely meant business, especially to warrant a gathering of this magnitude.

As he entered the building, LaDon took a quick look around. The seating reserved for the Lead Representatives had yet to be extended, and the teleportation pads were not yet active. *Something this huge should require their presence*, he thought as he rubbed shoulders with strangers, continuing his trek through the crowd.

He had already shoved his way through the thickest part of the masses, so he started his journey down the aisles. Behind him, the sea of Solayans continued to flood the entry points on all sides. He stepped up his pace, determined to find a seat in his favorite section in front of the Assembly before it was too late. He scanned the section for a seat. *Yeah!* He hurried along the aisle, sidestepping a few familiar faces, and plopped himself into the seat. *Ahh, yes. Perfect.*

LaDon's seat had just begun to warm up and he felt well nested in his spot. He scanned the room for any signs of the Assembly. Pomph's stories about the Assembly were always LaDon's favorites. LaDon watched intently, hoping to get a glimpse of Barton Urthorn, the Head of the Assembly who starred in all of Pomph's best stories.

Once, LaDon received two prestigious awards for enhancing the security and integrity of the archives, and had his picture taken with Barton. He remembered vividly watching the photographer wave for their attention. LaDon held one of his favorite Solaspheres balanced on a hip and Barton had one arm around him. LaDon had felt eight years old again. He was sure everyone could see him shaking with excitement. They could keep the awards, because that photograph was the personal highlight of his career. LaDon looked at this picture every day, proudly displayed on the wall in his office.

Down on the main floor directly in front of him, he noticed water dispensers in their rightful place atop the long, curved table extended for the Assembly. A few minutes passed, yet there was still no trace of their presence. He thought how nice it was going to be seeing the Assembly in person rather than through the inputs of his Solaspheres.

Day in and day out, LaDon watched recordings made by the Solaspheres. His job was to ensure the recordings were filed appropriately and to document anything significant about each one. Solaspheres were the tool of his trade. The spherical devices, just a little bigger than a Solayan's head, had a reflective band around the equator that recorded a full three-hundred-sixty degree field of vision. The outer shell acted as a giant sensor, recording every sensory input a Solayan brain could handle. It was the ultimate in virtual reality. But there was something to be said for seeing life through his own eyes.

With his mind drifting toward work, he recalled the last recording he viewed before coming to the Prime Hall. Earlier that day, he had notated a protest as worthy of review and took time to listen to some of their concerns. The recording had played a scene of a heavily congested parking area in front of a dilapidated, historic building. Fifteen to twenty peaceful protesters gathered, ready to speak their mind about the issue. Just before the Solasphere teleported back to the archives, an elderly lady approached. With one hand on her hip and the other supporting a waggling finger, she scolded the Solasphere.

"You do not have the right to demolish this building! This is a relic! It is a sign of our past, Solaya! Your Solaspheres could never capture the essence of a place such as this! Please, let our voices be heard!" The lady beckoned to the Solasphere as the viewing screen went blank, leaving silence in its wake.

Amidst the constant hum of voices and shifting of bodies in the Prime Hall, LaDon recalled how he felt while watching the recording. He admired the lady's spirit and the passion she showed for her cause. Still, no matter how much LaDon twisted his mind, he saw no purpose in her ranting and raving. It was just a building. Besides, the Solasphere recordings of the structure were properly archived and available to the public. Anyone could experience the building simply by visiting the historical archives. LaDon did not value material things. He had the luxury of his memories. He discontinued his search for understanding and chose simply to sympathize with her distress. He had filed the event under the peaceful protest section of the historical archives.

The murmur of voices in the Prime Hall intensified. LaDon's mind snapped back to the present. The influx of spectators had finally slowed, but something else was happening. There was a deep rumble high in the rafters. LaDon twisted in his seat and looked up. The restricted sections high above him on both sides of the Prime Hall were extending. This only meant one thing. The Lead Representatives of each nation had been invited to attend this event. The chatter in the room changed its cadence as different voices emerged that were once silent. The gravity of the moment took hold of LaDon. He felt as if he had just stepped out of a pool soaking wet. *Why are they here? This is big. No, this is huge. I better be on my toes. Oh wait, I think I might vomit.*

The unique deep whirring sound of the teleportation pads powering up filled the room as security personnel ushered the Lead Representatives to their pads. Solaspheres trailed close behind them like moths searching for a flame. Since he was responsible for documenting important political events, LaDon knew he would be watching again later. He counted each representative as they took to their pad for teleport. Their flowing robes were tied with cloth bands that represented their respective nations. Straining to see, LaDon glimpsed of the Lead Representative of the Vaknoreeyan people, LaDon's beloved nation. Joh Lin had been his representative for as long as he could remember. With Joh's wit, tough words, and sometimes abrasive personality, you were guaranteed an entertaining show.

Gradually the commotion following the entrance of the Lead Representatives started to diminish. LaDon turned his attention back to the seating area meant for the Assembly. At any moment, they would enter the Prime Hall and take their places. LaDon tried to focus on the arrival of the Assembly, but around him Solayans still talked amongst themselves, shaking hands with friends and motioning to others across the way to come and sit next to them. LaDon noticed faces from his office, but he was too excited to socialize. Some made eye contact and tried to acknowledge him, but they received no response. All of his friends knew not to take it personally when LaDon was focused on anything exciting. He was surprised everyone else seemed so calm. Selfishly, he resented their chattiness. His excitement was about to rattle him apart. *Maybe if everyone would just find a seat, I could relax,* LaDon thought.

Just as LaDon let his tension loosen, people began shifting in their seats causing a enough noise to get LaDon's attention. He whipped his neck toward the doorway closest to the Assembly's table. The door opened with a click. First to enter the room was Blaine Steele, Spokesman of the Assembly. LaDon imagined a smooth, bass-line rhythm as Blaine strolled across the floor with his usual swagger. Each month LaDon documented Blaine's addresses to Solaya. As Spokesman of the Assembly, Blaine was sure to say plenty today.

Behind Blaine followed Aleen Fabian. With her demure, ladylike gait, Aleen was the most recognizable. LaDon always admired her elegance and grace. Aleen's sophistication seemed so natural, watching her made you realize you were slouching in your seat. Although her posture and poise were at peak performance, she wore an expression of concern. He quickly searched for any memory of a Solasphere recording that might have warranted such a look. He came up blank.

After Aleen took her seat, LaDon saw another figure crossing the threshold of the entryway. *Is it him?* It wasn't. Instead, the likable Alex Cuberly walked through the door. Watching recordings of Alex made LaDon wish they could be friends. His energy drew you in. This feeling was somewhat humorous since Alex probably didn't even know LaDon's name.

There was still one Assembly member left to arrive. For LaDon, and probably plenty of other Solayans, this was the most important member of them all. His mind raced. The worry on Aleen's face started to make sense. *Where could he be? Why isn't he here yet? Has something happened to him? You don't think...*

"Order, order I say! the Assembly demands

order!" Blaine Steele shouted over the sound system as the sea of distressed tones filled the Prime Hall, swallowing Blaine's voice like a black hole.

Blaine looked younger than his years. With his square jaw, muscular build, jet-black hair, and clean-cut beard, he was known to receive plenty of attention from the ladies. As Blaine struggled for order, LaDon noticed his contorted expression and furled brow. Blaine slammed his gavel, but it proved useless. The spectators waved their hands while shouting over each other as Blaine attempted to gain control of the room.

One man's voice rose up above the roar, "Just what exactly does this new science do?"

"Which nation was the first to discover it?" said another voice just a few rows in front of LaDon.

LaDon turned in his seat and arched his back, straining to put faces with the voices. Those asking were higher ranking officials. *Wouldn't they know something by now? Are the higher-ups in the dark on this one as well?* Voices continued to ring out over the chaos as Blaine looked like he was about to give up hope.

One seat remained empty. LaDon watched Aleen. If anyone noticed the particular Assembly member missing, it was Aleen. Her aged face and pursed lips betrayed the worry she felt whenever Barton was not by her side. She had earned the respect of the entire Delnokeeyan nation due to her work on propulsion, as well as her keen political sense and way with words. Nevertheless, her work had come with a price. Time had done wonders for her career, but had taken its toll on her body. She had proudly earned the wrinkles on her face along with the salt sprinkled through her hair. Aleen often said during press releases that she would do it all over again as long as Barton was with her every step of the way.

LaDon tried to read lips as the Assembly members talked amongst themselves.

"Have you seen Barton?" Aleen whispered as she leaned across the empty seat to Blaine.

"No my dear, I have not," Blaine blurted accidentally into his mic.

He responded with a quick but courteous tone, still futility beating his gavel across the block.

"He'll be here, Aleen. Don't worry," Alex said with a confident gaze.

"Oh, I'm not worried, Alex. I just feel better when he's here." Aleen wrung her hands continuously with angst.

At this point, one would have expected the voices to diminish. Well, they didn't. More cries came from the masses.

"Assembly members, please tell us why this is being kept secret?" a woman from the crowd demanded.

Blaine tried to respond, "Jillian, we are not trying to keep anything a..."

But it was no use. Blaine dropped his gavel in defeat. LaDon could tell Blaine wasn't going to be heard over such commotion.

Finally a Solayan's voice a few rows in front of LaDon rang out above the rest, "Can we trust the

Assembly with something we are yet to understand?"

After the daring question, a lull rolled across the crowd like a cosmic wave. Aleen's face reflected a change in the atmosphere and explained the sudden hush of the voices. It was as if she could sense him. All worry evaporated from her face. LaDon imagined the presence of his soul surrounded Aleen like a cloak of peace amongst the discord. The crowd recognized his presence as well. The lull softened more and eventually became silence. It was as if order had a name, and that name was Barton Urthorn.

The first thing anyone noticed about Barton was his respectable stature towering over the masses along with his broad shoulders. With his white hair and weather worn skin, his face told an epic story of love and hate, stirred in a cauldron, and topped off with heartache and betrayal. Barton took his seat at the center of the table and arranged his robe. He gently caressed Aleen's hand. She was complete again. Through his war-torn façade, he smiled at her with unmistakable true love.

As far as LaDon was concerned, the meeting could truly begin. Barton faced the masses with calm confidence. In the same voice that had calmed Solayans for many years, he responded with a slow, even tone that captivated its audience once again.

"Who would ask such a preposterous question? Just why do you think we are having this meeting? I can guarantee you it's not for my health," he said with a grin, which brought smirks and even chuckles to previously crabby Solayans.

LaDon joined in the laughter. Barton's witticisms may have been rare but were always ingeniously executed when needed. The overwhelming tension settled, allowing LaDon to clear his mind. He paid close attention to the body language between the Assembly members. Blaine looked toward Barton. With a reassuring nod, Blaine immediately began his duty as Spokesman of the Assembly.

Chapter 2

Technicalities

"Ladies and Gentlemen, I would like to call this meeting to order. We thank you for attending such a momentous gathering. We apologize for the short notice on our part. The meeting agenda, along with a small synopsis, will now be broadcast to you. Please be respectful of others and do not read aloud. Hold all commotion, noise, and emotional outbursts until your viewer goes dark. Now, please tune your optical viewers to Wave Four to receive the transmission," Blaine explained in his usual business-like tone.

The crowd hurriedly adjusted their viewers to receive Wave Four of the Planetary Optical Viewing Network, or POVN. Waves one through five were restricted to planetary satellite up-links, designated for local news, emergency response, and special events. LaDon focused his eyes intently on his heads-up-display just in front of him. He was determined not to miss a word.

LaDon had always admired the bioengineering skills of the Calereeyan nation. The Solayan people had come to rely on the viewer for everyday use. For LaDon, it was a necessity. Without the implant, he would not be able to interact with his Solaspheres.

Blaine continued, "We have prepared a small bit of information. Please read along with me.

Afterwards, we will take time to answer any questions you might have. Let's begin, shall we?"

LaDon sat up in his seat and began to bounce his leg. This gave him a small outlet for his excitement. With all the inventions discovered to date, LaDon couldn't imagine what was about to appear in his viewer. *Another advancement in space travel? A more effective matter replication method?* This was the first official release of any technical documentation since the announcement of the meeting. The audience was about to get answers to the questions they had brought with them to the meeting.

LaDon watched intently as his viewer showed the documentation. LaDon listened as Blaine read aloud:

Synopsis of Gathering

Three months ago, a Delnokeeyan scientist by the name of Jendall Kimnor approached the Assembly to propose a theory. To his surprise, the Assembly had recently met with a Kalloneeyan scientist just weeks before named Phelix Castor. It was decided to have Jendall and Phelix meet, share their ideas, and see what might come of the meeting.

After one month, Jendall and Phelix solved their riddle. It was as if Jendall built the lock and Phelix crafted the key. the Assembly immediately took action to assure each race got equal credit in the discovery.

* * *

Blaine paused for a moment. LaDon focused away from his viewer. He watched as Blaine looked across the crowd. With everyone glued to their viewers, Blaine continued reading.

"Find 731 of the Planetary Scientific Consortium gives us another historic find in Solayan history. With equal collaboration of two of Solaya's finest scientific minds, Jendall Kimnor of the Delnokeeyan Community and Phelix Castor of the Kalloneeyan Community, we, the Assembly, are proud to present this fine astrological achievement: a new breakthrough in interspatial travel which combines spatial displacement theory and vibration theory. In layman terms, we now have the technology to explore a new realm of existence which is closer to home than our own neighboring planets."

All viewers went blank, and the following message appeared:

"More information in a moment..."

You could almost feel the intensity as everyone stared at blank viewers and began to process the vastness of such a profound discovery. Loud cheers erupted, especially from the Kalloneeyans and Delnokeeyans. Physical embraces and shouts of congratulations took place between the nations. Other nations directed their applause toward the Delnokeeyans and Kalloneeyans as they showed their support for such a momentous but still puzzling find.

After giving the crowd a few minutes to digest the information and converse amongst themselves, Blaine once again looked to Barton for the authority. With the go ahead, Blaine firmly requested order, and the masses settled into tense silence, fully focused on the next words from the Assembly. They were prepared to receive the full scientific description on their viewers.

LaDon was speechless. It was exactly the kind of news he had hoped. A new find. A new discovery.

LaDon sent a brief message to Jendall. They had worked together in the past when a particular Solasphere recording documented Jendall's work. His message queue was probably already full of congratulations from friends, family, and even strangers.

Blaine transmitted the next page of documentation. "As you can now see in your viewers, I am displaying the full agenda of this meeting. There's a lot of information here, Solayans. We will be revealing the information in sections. We don't want anyone reading ahead. You might miss important information we wish to relay that is not documented here. First, let's review the agenda of this meeting."

LaDon always hated the formality of meetings. It slowed progress. *It's a meeting. We definitely can't forget to read the agenda,* he thought sarcastically.

"The first item, as we are all aware, is the reading of the summary. Obviously, we have already read the summary. We placed that in front of you prior to showing the actual agenda so everyone would understand the seriousness of this meeting. Next on the agenda, we will cover the technical aspects of the science behind this extraordinary find. Following that, we will explain what stage we are at in our research. Next we will discuss appropriate monitoring and policing of this technology. Finally, we will take questions in an orderly fashion via the booths at the end of each aisle. When your viewer reads "Questions", the nominated speakers of each nation will make their way to their respective booths." Blaine said as he adjusted himself in his seat as he finished.

It was almost as if the crowd was sitting on a bomb that might explode under their seats at any moment. Blaine took a sip of water. LaDon was reflecting on Blaine's words, anxiously waiting for the next syllables to fall from his lips. He knew that anything worth hearing would come from Blaine himself. With his ears finely tuned, LaDon started to hear small voices whispering about those last comments.

"Police? Why would it need to be guarded?" someone whispered a couple of rows back.

"Maybe it's dangerous," another voice responded to the question.

"Oh don't jump to conclusions," someone else scolded from the general direction of the conversation.

"Blaine said there's a lot more information still to come. Give the Assembly a chance to explain. They seem to be doing well thus far," reasoned a female voice just one row behind LaDon.

That voice of reason back there seems correct. LaDon tried not to lose focus on what Blaine would say next. *Right, just give them a chance to explain. It's their job. I wish they would all just be quiet. No one cares what they have to say anyway.* Just as those thoughts left his mind, LaDon could see Blaine shift in his seat and his facial demeanor change signifying he was about to begin speaking again.

"Please remember, we ask for no commentary during the reading. Let's keep order and be respectful of each other as we delve into the heart of the information. If you miss something, don't worry. We have dispatched multiple Solaspheres throughout the building to cover the event from all angles. This meeting will be public record once it is adjourned. Just sit back, relax, and we will get through this," Blaine said as his focus dipped into his viewer and he began to read the material aloud to the audience.

With the combination of Delnokeeyan matter relocation and the Kalloneeyan theory of multiple universes, we as Solayans can now relocate objects from a point in space-time in our universe to a point in space-time in another universe. What at first was believed to be time travel within our own universe proved inaccurate with testing. Instead, we can send an object into a neighboring universe, record the immediate surroundings, and bring the device back to Solaya. We have documented evidence showing the existence of another universe, which seems quite like our own. However, at this point in our planet's history, we do not understand all the mathematical variables which dangle in our equations.

Technical Discussion Journal for Find 731

Using the Delnokeeyan Solasphere (ref. Planetary Scientific Consortium - Find 073), we have relocated numerous devices, in fifteen minute intervals, into a separate universe and brought them back safely. Upon return, the Solasphere is attached to a terminal and the data is uploaded into the archives.

Ninety-five percent of all devices have successfully returned. We do not understand the disappearance of the missing five percent, but theories are being tested to find out why these devices did not return. We have attempted to map stars as we try to navigate our way through the newly found universe. This has proved difficult. We cannot seem to find any consistencies in star patterns to derive a coordinate system.

As of the date of this publication, our team of scientists are now focusing their attention on one issue. In the equations that power this relocation, there is an inconsistent variable. No matter what value is given, the device still reaches its destination and returns to Solaya. Without the variable, the device goes nowhere.

A small roar of voices arose as Blaine paused before continuing into the next section. LaDon was in complete shock. His heart began to pound. *Who is currently reviewing the recordings? Where exactly are they sending the Solaspheres? What exactly are the devices recording? Another universe? Did he hear that correctly? He didn't say another galaxy. He said another universe. How is this possible?*

On the Assembly stage, Alex slouched in his seat with his hips shifted to one side. He was propped on one elbow while rubbing his face with his opposite hand. In classic Alex Cuberly style, the look on his face was somewhat nonchalant. Aleen was sitting upright, like normal, being the lady she had trained herself to be. She had a tiny look of worry on her face, but this could have simply been from unrest among the sea of spectators filling the Prime Hall. Upon second glance, she seemed calm, for Aleen anyways. LaDon's eyes finally made it to Barton.

At that moment, LaDon's heart nearly stopped beating. He wondered it if he might faint. His mouth went dry, his breathing stopped, and his throat began to rival that of dry desert sands. Barton Urthorn was looking directly at him! But how could he be certain? LaDon averted his eyes and tried to convince himself he did not see what he just saw. He looked to Barton once more. There was no denying it. He was certain that Barton's gaze never faltered. Barton Urthorn was, in fact, looking directly at him.

He wondered if there was anything he could have done to keep his birth from ever happening. Nothing came to mind. A small beep in his viewer drew his attention away from Barton's gaze. It was an incoming message from none other than Barton

Urthorn. LaDon's mind was stuck on the message flashing in his viewer. The decision to open it lasted less than a second. With a quick swipe in the air, he quickly opened the message. Skipping the formalities on the first few lines, he saw one simple line within the body of the message.

"LaDon Grafter, please report immediately to Nalkalin after the conclusion of this gathering."

Me, go to the home of the Assembly? LaDon read the words over and over. Finally, he closed the message on his viewer and slipped back into reality long enough to see Barton drop his gaze and turn his attention back to Blaine. The next few sections were a blur. Blaine's words might as well have been inside a soundproof box with no immediate means of escape. LaDon did his best to focus on two things at once as Blaine continued reading.

.

What We Know / What We Do Not Know:

- *We know that Solaspheres are traveling into a separate universe.*

- *We do not know how to accurately access where the devices end up.*

- *We know there is an unknown variable in our equations. Without it, the science fails.*

- *We do not know why only ninety-five percent of all Solaspheres return.*

Policing the Find

the Assembly has decided that in order to police this new find, there will be a new enforcement division created. This division will be known as the Galactic Research Committee (GRC). The GRC will report all of its findings to the Assembly, where any issues will be discussed, and the Assembly will reconcile any unresolved matters with the assistance of the Lead Representatives of each nation. This division will be devoted to the understanding, development, and overall future of this new science. To ensure those chosen for these positions are knowledgeable in the science of this new find, the candidates will be selected strictly by the Assembly.

The next screen displayed a black background with bold, white text reading *Questions*. This text remained for approximately five seconds. Finally, all viewers went blank. The transmission had ended. For only a moment, the sound within the Prime Hall evaporated like a soul leaving a dying body after its last breath.

If someone could look inside the actual heart and soul of every Solayan at this moment, one would see a hodgepodge of different emotions ranging from alarm, disbelief, pride, but most of all, confusion. Suddenly the silence was filled with the whir of the teleportation pads. The attention of the congregation was diverted. Everyone turned to watch as the representatives from each nation made their way to the designated booths.

LaDon's attention was not so easily diverted. He felt as if he was about to stand up and give a speech about something he knew absolutely nothing about and wearing only his underpants. He was slowly accepting that he had just received orders to report directly to Nalkalin, the home of the Assembly. He had no time to prepare. The message strictly said to report immediately after this meeting. *That had to be why he was staring; why else would he make such deliberate eye contact with me? Then again, why make such eye contact with me at all? Just because I was about to receive a message from him? It doesn't make sense.* LaDon was certain of one thing. No one was asked to report directly to Nalkalin unless you were a Lead Representative or had vital scientific information to relay. LaDon was nowhere close to being a Lead Representative and science was only something he watched through the eyes of his Solaspheres. *Ah, the Solaspheres! That MUST be it. But shouldn't they contact a Delnokeeyan scientist? They built the things after all.* All of these ideas bum rushed LaDon's head faster than he could come up with new ones.

LaDon took advantage of the shuffling of delegates and attempted to relax himself. He loved Nalkalin. Another monstrous, elaborate structure full of Nuweeyan innovation, Nalkalin was known for its fortification as well as its size. It served many functions for its residents. Among other things, it served as the Assembly's meeting place and living quarters. Its vastness and beauty had been captured by numerous Solaspheres. LaDon had viewed a majority of these recordings. Many of the recordings covered the Assembly doing what they do best.

Quite literally, the Assembly was treated as royalty within the walls of Nalkalin. As Barton had pointed out in past news conferences, the Assembly looked forward to peace and quiet since the life of an Assembly member is always on the go. Each Assembly member was assigned separate quarters and a personal assistant. The personal assistant's job was to keep each member attuned to the lifestyle of the Solayan community as well as assist with scheduling and time management.

LaDon thought about this for a moment. His message was addressed directly from Barton Urthorn and not his assistant. LaDon scanned the room and located Barton's assistant, Codi Sphin, who was still active. *What does it mean that Barton sent this message directly without the aid of his assistant?* This thought brought LaDon rushing back to the same level of tension that still filled the room around him. He realized he was going to have to stay in that state until he reached Nalkalin and could talk to Barton face to face. Head to head with the man he admired most in the world, next to his grandfather Pomph, of course.

Chapter 3

Dare to Question

The Lead Representatives and their entourages began making their way down to the main floor. Talk of variables, shifting matter, and other universes filled LaDon's ears like he was being submerged in water. At some point, it became a blur of noise. He lost the ability to distinguish individual conversations. His mind and heart raced as the representatives hurried toward their respective booths. Each nation had to trust their elected leaders to ask the right questions to successfully represent them. At last, all representatives reached their respective booths and were ready to begin the inquiries.

LaDon could see by the look on Blaine's face that he was prepared for the onslaught. This was what his role as Speaker of the Assembly was all about. Blaine knew the Assembly would be under fire and must be prepared.

The first group to ring in was the Lead Representative of the Delnokeeyan nation, Ryndal Jomere. Ryndal's face was twisted in thought. LaDon could tell Ryndal had questions that demanded answers. Ryndal must have already formulated some interesting questions as he stepped up to the podium in haste.

"Greetings, Assembly members. I guess I'll get

right to it. Just to get this question out of the way, how long has the Assembly known about the existence of this science?" Ryndal asked in his firm, business-like voice that had represented the Delnokeeyans for the past thirty years.

Blaine responded without hesitation, "The exact discovery date was twelve days ago, Sardis 16, 6476. This is this official date the science was announced to the Assembly. As for when Jendall and Phelix actually discovered the science, the exact time is unknown to the Assembly at this time. Feel free to ask Jendall or Phelix directly."

Immediately after Blaine's statement, Alex cut in with a somewhat firm tone, "Also, if I may add, once we heard this news from Jendall and Phelix, we immediately split them into separate living areas and quarantined the research facility until we were able to fully understand what had been discovered. We understand this science must be kept above accusation until we can properly police its behavior. We are taking every precaution to assure this finding is well documented and preserved."

Due to LaDon's Solasphere reviewing, he could tell that Blaine was somewhat relieved that a Delnokeeyan representative had asked the question. After all, it was a Delnokeeyan who had discovered the science. To question another nation's morality or ethics at such a sensitive moment might have caused an all-out uproar in the Prime Hall.

The light dimmed on Ryndal's booth. The next light to register came from the booth occupied by the Lead Representative of the Vaknoreeyan nation, Joh

Lin. LaDon sat up and moved himself to the edge of his seat to show respect. He also needed a better line of sight between Joh and the Assembly. Joh was well-known for bringing his best hand to any gathering. LaDon admired the gall and candor that surrounded Joh. Sometimes it took rash questioning to get to the bottom of things. No need to maneuver around the feelings of others. Strike at the truth.

Even though Blaine would probably be answering the question, LaDon knew Barton had his back. He always did. Barton was Joh's sharpest adversary in this type of setting. They balanced one another quite nicely. Wit and reasoning versus rash thoughts and implications. As Joh prepared his notes to begin speaking, LaDon imagined the two politicians in a small-scale, virtual war of the minds. He visualized a scene of Barton and Joh in a skirmish. Swinging their swords with graceful technique as they went round and round, battling for the truth to be properly delivered and fully disclosed.

With his usual smug look, Joh began, "I will go ahead and be the first to point out the obvious. There has been a lot of information to absorb after reading this well-written documentation covering Find 731. Although, it seems to me there are multiple discoveries wrapped into one. With that said, I have two questions if that is acceptable?"

Joh directed his question directly to Blaine. As expected, Joh was off to a roaring start, pushing the limits of the proceeding. Blaine leaned over to look in Barton's direction, followed by a sweeping look over the entire Assembly. Any action in a gathering that was outside normal procedure must be approved by Barton. All Assembly members had the right to deny the request, but the ultimate word came from Barton. Blaine and Barton's eyes met. Barton nodded firmly, allowing the request.

Joh Lin continued, "I thank you graciously. My first question for the Assembly: Can we derive from this documentation that we have discovered, not only a new universe, totally separate from our own, but we can also pass objects back and forth, to and from, this new plane of existence?"

Blaine started to answer, but Barton sat up and placed his hand on Blaine's shoulder. Blaine stopped and looked calmly in Barton's direction. With a knowing expression, Barton leaned forward, propping his elbow on the table.

With a smirk, Barton responded, "Well, Joh Lin, you ask if there's a new universe and you ask if we can pass objects over there. If my ears serve me well, that sounds like two questions wrapped into one, if I can point out the obvious."

The verbal jab at the end of the sentence caused the crowd to respond with a bellow of laughter, including Joh Lin. Joh leaned on the podium and looked down with a defeated grin on his face. Barton remained forward in his chair, ready for the electricity of laughter to settle. This exchange seemed to ease the tension in the air. It felt like the crowd had been waiting to hear Barton's voice since the beginning of the gathering. It was a much-needed relief. It was just what the room needed to calm the tension without disturbing the seriousness.

As the laughter subsided, Barton continued, "Now, to answer your first question, Joh Lin, yes. There are, in fact, two discoveries here, but we are keeping them under one documented find as they were discovered in tandem. The same nations were there upon discovery. The first time the device disappeared from our sight, it traveled into the new universe; therefore, they discovered its existence. the Assembly discussed this very thing in length. We came to the conclusion to leave it as one find. This is one reason I feel this find is such a momentous occasion for our planet."

Barton gracefully stood from his seat and moved to the front of the table. LaDon felt as if he was watching a performance. A form of entertainment, with Barton as the main attraction. The way the light moved across the platform, his calm gait, and the intense emotion of the crowd all meshed together like a symphonic orchestra.

"You see, our history has shown that discoveries such as this can unify our nations even more so than we are today. Also, on the darker side of the story, such discoveries have almost unraveled everything we have fought so hard to create and maintain. It is what we decide to do with this technology that will tell us if we are strong enough and capable enough to handle something of this magnitude. It is up to you, Solaya, to understand and embrace this new science. Should we be afraid? Yes! Should we be mindful? Yes! Should we be hopeful? Of course! Just be assured of one thing. Everyone you see standing at their respective booths were chosen by each and every one of you. Each person at this Assembly table was a unanimous decision by those elected officials. We have been through thick and thin together. Finding after finding, unifying our thoughts, making us stronger and more efficient. I promise you, as your Assembly, we will not change. This method of thinking works. It's worked for decades and it will continue to work if we let it. We are treating this as a matter of utmost urgency and caution. But like every other science we've uncovered, it takes time to learn. It takes time to get it right. So I ask you, people of Solaya, be patient but be vigilant." A rumble of applause followed Barton's final thought.

Reaching his seat, Barton looked toward Joh Lin and nodded. "You may continue with your questions, dear sir."

Barton's way with words was legendary, but one never fully appreciated the wisdom until it slapped them on the face. The planet of Solaya needed assurance at such a questionable time as this. After hearing his words, everyone knew Barton knew it as well.

Joh Lin continued, "You can trust the Vaknoreeyan people to fully support the Assembly. We will do our best to aid in the enrichment of this new discovery in any way we can."

Out of respect, Joh Lin paused to allow Barton or any of the Assembly members to respond.

When there was no response, Joh Lin continued, "My second question concerns this unknown variable. Are there any leads in understanding of its function within the equations?"

Looking to Barton out of respect, Blaine replied, "We have allowed Jendall and Phelix to continue their work as of this morning; therefore, you know what we know up to this point. Without the variable, the science is useless. With the variable set to zero or to any negative number, we get the same result—nothing. They never move. If you set the variable to anything else, you get a result. Just like with every find, the team will report any updates to the Lead Representatives. Public releases will follow as we gather answers to these types of questions."

The question and answer session continued for the next few hours. Each nation got a chance to ask as many questions as they could muster. Each representative wanted to make sure they satisfied their nation as well as their own curiosity. The focus of the questions stayed on topic for the next dozen inquiries. Questions ranged from concerning the information the Solaspheres actually recorded to whether or not there were any photos or videos scheduled to be released. The questions concerning Solasphere data particularly interested LaDon. He moved to the edge of his seat again for the next question from Sandren Dray, as she had proved to ask the most questions concerning the recordings.

Sandren began, "Can one of you comment on the senses the Solaspheres are picking up during their journey?"

Aleen leaned in to respond, "The recordings pick up freezing temperatures. It tastes nothing. No sound is recorded. It records no scent. We believe this is simply because the device is in the vacuum of space. Visually, it records what appear to be stars in the distance. We have yet to land any of the devices close enough to these objects. In fact, that's one of the problems. We cannot predict where it will go. I wish I could tell you more, madam."

LaDon agreed with the assumption that the device probably landed in deep space. Cold, nothing to taste, no smells, no sound, and it could see stars in the distance. He imagined it would look just like his back was to Solaya and he was looking out into the cosmos. *Why can't they predict where it goes?*

LaDon noticed Barton looking at Blaine. Barton signaled for the final question of the evening, to be presented by none other than Joh Lin. All bets were off and things just might take a turn in a different direction. Joh Lin did have a flair for the dramatic.

After Joh Lin was recognized by Blaine as being the last question of the evening, Joh Lin began to speak, "Assembly members, we have listened to numerous questions from our brothers and sisters. We have questioned the date and time of the discovery. We have questioned placement of due credit. After hearing the results from the Solaspheres, we have questioned what preliminary findings this new discovery has brought forth. We have even asked about the actual science supporting this find. I believe I can speak for everyone present in saying that we have received answers in great detail for each of these questions as well as others."

LaDon could feel the masses becoming a bit on edge. When Joh Lin recapped his thoughts, not to mention having the final question of the evening, he was always preparing for a tough question. This was Joh Lin, after all. He was the Vaknoreeyan famous for his deep thought, tough love, and talent for putting his people first. It is a big part of what got him elected.

LaDon could see Blaine's uneasiness, not anger, but frustration at Joh Lin for taking his sweet time to paint an elaborate picture before asking his question. Even Barton shifted a bit in his seat while still keeping a placid smile upon his face. Aleen, of course, looked as stern as a librarian. She had no problem displaying her feelings for others to see. Alex Cuberly wore his normal poker face. He could fool a room into thinking he was asleep with his eyes open.

LaDon had watched plenty of interviews with Alex. LaDon knew Alex was probably thinking something quite severe toward Joh Lin. LaDon imagined what theater production could possibly be playing inside Alex's head. He was probably imagining the intensity of Joh Lin's screams as he pushed Joh Lin from a space craft hovering at the edge of the atmosphere. This brought a tiny smirk to the corner of LaDon's mouth. LaDon watched closely as Barton and Alex's eyes met for a brief moment. Barton grinned as if to let Alex know he could read his mind. It was as if all of them, including LaDon, were in on the joke.

The entire Assembly listened as Joh Lin finally reached his point, "I understand this is the final question of the evening. I guess I better make it good. This is for all members of the Assembly. Have there been any attempts to transport living specimens into this neighboring universe?"

The mere thought of such travel for a living being had never been imagined. To travel as a Solasphere, and into another galaxy no less, was unheard of. Joh Lin knew this of course. This caused the lull of the crowd to become more intense. Joh Lin quietly took his seat looking slightly smug. He knew what he was doing. Without hesitation, Barton stood to answer such a daring question.

"No," said Barton, with a short, brisk tone.

LaDon knew that was all Barton had to say. As Joh Lin intended, this question put thoughts in everyone's head, including LaDon's. The endless possibilities raced through his mind. *What if they do find a star system? What if it has a planet? What if it has life? All of the current work on deep space travel that Solaya is working so hard to conquer could take an immediate shift.* So many scenarios formulated in LaDon's head as he pondered the vastness of such a statement. With Barton's answer of no, everyone could only speculate as to the magnitude of this new discovery.

At one glance from Barton, the Assembly, except for Blaine, rose from their seats and began exiting through the side doors of the Prime Hall. Generally, LaDon would have expected to receive a detailed elaboration on that no. Barton obviously had no time for a flowery speech. The question was answered clean and direct. Plus, it *was* the last question of the evening. Many Solayans began to move towards the exits. Over the bustle, other questions pierced the bedlam in the chaos as the Assembly exited the Prime Hall.

"Has this concept even been fathomed up to this point?" asked one voice.

"Are there plans to test such a thing?" yelled another.

With the audio system at full volume, Blaine's voice rang out with clarity, "Thank you all for attending. the Assembly will be giving updates on this matter at regular intervals throughout the rest of the year. If you wish to communicate with us concerning this matter, please have your representatives visit the Assembly's headquarters at Nalkalin and schedule a date and time for your inquiries. Please keep in mind that our time will be limited as we explore this matter thoroughly."

Blaine exited as well. LaDon knew that the Assembly's next duty was to convene at the wondrous place known as Nalkalin.

This thrust LaDon's mind to the order he received from Barton. The sinking feeling in his gut returned. His nerves began to sizzle again. This meant he must follow the Assembly to Nalkalin. *I wonder if anyone else received any orders during the meeting?*

People went their separate ways at the Prime Hall exit. Some ventured back toward their places of work. Some seemed to be heading to their homes. Others seemed to be walking to their transports. The crowd started to thin around LaDon's path toward

Nalkalin. One by one, people drifted off the path toward other destinations. Suddenly, he was alone. No one was behind him. No one in front of him. Not even the Assembly members were in sight.

Maybe they took a transport.

The walk to Nalkalin was not that far, and LaDon didn't mind the exercise. It would also give him time to think through everything he just heard. Thoughts of deep space travel made him think of his parents. *Grandfather would've loved to have heard this.* LaDon smiled. He imagined some choice phrases that Pomph might have said in this situation. He could still replicate his voice in his head any time he wished. As his emotions ran deep, LaDon's eye let loose a small tear. He was not sure if the tear was sadness for Pomph, or self-pity for having to live this moment without him. In any case, there would probably be many more findings where this one came from. He would have to make his own way while keeping Pomph's memory alive.

LaDon started to notice his surroundings. He realized how far he had traveled while his mind was adrift and filled with the thoughts of love and home. He was now standing in front of the place that housed the governing body of his entire world, a place he'd seen often through the eyes of a Solasphere. The only building that LaDon felt was worthy of admiration. Nalkalin. If only Pomph could see him now. He would have been the proudest grandfather on the planet.

Chapter 4

Just What Do We Have Here?

LaDon approached the main entrance. The security guard greeted him immediately. LaDon recognized the security guard from a recent Solasphere recording he had archived just a few days ago. The guard gave him a little more trouble than he was used to when traveling by Solasphere.

"Greetings, fellow Solayan. Do you have an arrangement with the Assembly?"

By the sound of his voice, you might say this guard couldn't protect much of anything. Once you saw his hefty stature and muscular build, you'd think otherwise.

"Yes. Yes I do. I received a message on my viewer during the meeting just now in the Prime Hall," LaDon replied courteously.

LaDon felt happy that this security guard was on his team. He also felt safe that he was invited by none other than Barton Urthorn. Then LaDon realized he had no way to prove it without sending the invitation to the security guard. He had no physical piece of documentation, no badge, nothing.

"One moment," the guard said as he accessed his viewer.

What was probably thirty seconds in real time felt like a trip to the next planet over and back.

What is he looking for? Was the message I received a mistake? No, it said my name, specifically. During the meeting, Barton was looking directly at me, right? I still have the message in my viewer. I could sent it to the guard. No wait, can I? Is that allowed? Is the message confidential?

LaDon began a slight panic as his heart started to race. He imagined putting on a production of a well-planned fainting spell. Suddenly, the security guard began to speak, but not directly to LaDon. The guard seemed to be looking off into the distance. LaDon realized the guard was lost within his viewer.

"What's that, sir? Yes sir, Mr. Urthorn. Right away, sir," the security guard barked as if he was back in his academy days. "I apologize. I was unaware of your invitation. You may proceed through the gate, Mr. Grafter. Mr. Urthorn is waiting."

The message was real. He didn't imagine the whole thing, although deep down he wished it was all a dream. The guard had said Barton was waiting. *Is he watching me on the security cameras right now? Did he see the look of worry on my face? Did I make a strange face?*

It then dawned on LaDon that the security guard did not know he was coming. *Why wasn't I announced, especially to security? Everyone is probably suspicious now,* LaDon worried as the guard shuffled through some equipment. LaDon saw that the guard was preparing LaDon's badge. He immediately heard Pomph's voice echo in his head, *Wow, you would worry about absolutely nothing if it were possible, wouldn't you?* This helped him relax a little as the security guard passed LaDon a visitor's badge already populated with his information.

Ah, a Holopass. This was one of LaDon's favorite finds. A brilliant Caleryeean design which stored and displayed the information of the individual, including picture identification. The device displayed the information in brilliant resolution when worn. LaDon knew not to lose this badge. Security issues are not taken lightly in Nalkalin. With the Holopass lit up with the brilliant display of his information, LaDon fastened it to his right shoulder as proper etiquette dictated and continued up the path to the main entrance.

LaDon took a moment to admire the view and the luxurious surroundings as he wandered through the courtyard of Nalkalin. In all its splendor, Nalkalin took LaDon's breath away every time he visited. The courtyard alone, with its beautiful assortment of rare flowers and well tended grounds, was enough to please the eye for hours. He was familiar with all of the structures and smaller surrounding buildings, with their pointed tops and rounded off entrances. One particular area stood out above the rest. There seemed to be a new building. This was something he hadn't noticed in recent Solasphere recordings. *This must be new. It's so...round.*

LaDon had only seen Nalkalin through the sensory input of a Solasphere. This made him realize the accuracy of a Solasphere's sensory inputs. Every smell, every sight, and even the way the sun hit his face. The sounds of watering devices irrigating the grounds and the songs of birds which seemed to repeat a never ending lullaby. LaDon heard one distinct difference that seemed to drown out everything else. Compared to a Solasphere recording, this sound was new. The sound of his own movements. Click, clack, click, clack. His steps seemed to drown out all of the beauty surrounding him. He tried to tune out the sound and soak in the beauty once again. He felt the pull between both worlds. Real life versus a Solasphere recording. Between the sounds of his own garments swishing together and his footsteps pounding the pavement, he realized he could not enjoy the serenity of the grounds like he did through a Solasphere.

"I need to get out more," LaDon whispered aloud to himself.

LaDon reached the main entrance. Again he was met by a guard, who checked the Holopass on LaDon's shoulder. The guard's attention drifted into the confines of his viewer and then back to LaDon.

"Second floor, room one. The door is labeled The Unification Chamber," the guard instructed as he motioned toward the lifts.

"Thank you." LaDon walked in the direction of the lifts leading to the second floor, thus proving he was no stranger to his surroundings.

This security guard has knowledge of my existence. News obviously travels fast in Nalkalin, LaDon said while huffing under his breath. As for the Unification Chamber, LaDon knew exactly where this room was located. Not to mention the history behind it.

Many world-changing events had occurred there, and many Solaspheres had traveled this path leading to this historic room. He imagined it was the very room Phelix and Jendall met recently to discuss the latest find. That particular Solasphere recording had yet to be released.

Wow, won't that be a good one?

Knowing the historic value of such a meeting place, once again, LaDon began to fret. Heavily. His worry escalated quickly because he had been so distracted by all of the sights, sounds, and smells, of Nalkalin. He had not stopped to give any thought as to what the Assembly wanted with a simple historian such as himself.

Needless to say, LaDon was not just any historian. He had made many breakthroughs in data storage. He even made a few Solasphere tweaks. But he was too modest to deem himself of any importance. He was simple, and he liked it that way.

His hands began to perspire, the muscles around his shoulders tightened, and his breathing picked up as the lift came to a stop. He disembarked onto the second floor. He started down the long hallway, which ended with the door to the Unification Chamber. It looked just like the last Solasphere recording of this room. Barton was probably right behind the door, as well as Aleen, Blaine, and Alex. the Assembly, right there, just a few steps away. He reached the doorway. The cold metal handle warmed as he wrapped his hand tightly around the lever. He pushed the lever downward and eased his way into the room. The door opened and clicked closed behind him. When LaDon entered this room in a Solasphere, no one paid him much attention. Not this time. The eyes of each Assembly member were locked on LaDon. He took a few steps into the room.

Blaine stood up, smiled like a gracious host, and stretched his arm toward the chair in front of their desk. "Ah, Mr. Grafter, please, come in. Have a seat."

"Thank you sir." LaDon smiled and moved toward his seat.

The walk from the door to the chair felt like a mile. Though his nerves were rattled, LaDon tried to control his angst. He began to imagine he was actually inside a Solasphere recording in an effort to corral his nerves. This helped a little as he scanned the facial expressions of each Assembly member.

Blaine was all smiles as he reclaimed his seat. Aleen sat graciously with a pleasant smile as if trying to relax LaDon. It must have been obvious that he was nervous. Alex was sitting square with the table with his hands folded. His elbows propped the rest of his body in an upright position. Of course Alex's face was a little harder to read. LaDon tried to pin an emotion to Alex's expression. It resembled curiosity. LaDon's eyes finally moved to Barton. Sitting in the middle of the table, Barton could be seen easiest of them all. Barton was arranged in his favorite sitting position: leaned to one side, propping up on one elbow, and looking at LaDon just over the front of his right shoulder. His smile was reassuring, relaxed, but ready for business. This welcome gave LaDon time to slow his heart rate, focus his mind, take a deep breath, and ready himself for what was to come.

"Mr. Grafter, thank you so much for coming on such short notice. We, the Assembly, want to welcome you to Nalkalin. We want you to feel relaxed and welcome while you are here. From what we understand, you have made many visits to this facility through the eyes of our Solaspheres. You are probably already familiar with your surroundings." Blaine's voice echoed in that oh-so familiar tone.

"Yes sir, Mr. Steele, I have. On many occasions." LaDon's voice cracked as he tried to clear his throat.

"My dear, please do not be nervous or whatever you may be feeling at this moment. You look as if you are about to run out of the room screaming if one of us so much as sneezes. The only person that you should be skittish around is Alex, and he's just a little thing." Aleen gestured toward Alex with a calm smile on her face.

Alex dropped his posture and twisted toward Aleen with a shocked, innocent look on his face. Alex returned to his normal posture and glared innocently at LaDon.

Aleen continued, "This is no interview. This is no test. Although what we have called you here for today is very important. You must understand, we need you at your best, and being nervous because you're in front of us will only defeat the purpose of why you are here."

LaDon   breathed easier at her comforting words. He took a few needed breaths and relaxed the tension throughout his body. Feeling the blood rush back to his face, LaDon noticed the Assembly members themselves relaxed a bit. All members except Barton. Barton's demeanor seemed placid, but his gaze had yet to falter. There was business to attend to, and LaDon could see it in Barton's eyes. He had seen that look many, many times before. After studying Barton's gaze for a quick moment, LaDon turned his attention to Blaine.

"Aleen, I think that did the trick," Blaine said. "Now, I would like to take the time to introduce each Assembly member. I feel if you got to know us a little you would..." Blaine trailed off as Barton stood from his seat.

"Mr. Grafter.", Barton said LaDon's name slowly, like a lawyer addressing a courtroom of jurors.

Barton nodded in Blaine's direction with a courteous gesture to excuse the interruption.

"Mr. Grafter, there is no need for introductions. I believe you already know us. Based on your current occupational status, and if my intuition serves me, you might even know us a little better than we know ourselves," Barton explained as he smiled at LaDon.

"I'm not sure about that last part sir, but I would agree with most of your assessment." LaDon reflected Barton's formally respectful demeanor.

"Good, with that out of the way, we can get

down to business. Your work with Solaspheres is unparalleled. No one on this planet knows these devices better than you. You know how they work. You understand their limitations. Most of all, young man, your keen observational skills are precise and even imaginative in some cases. Who else would have noticed that the announcement of Find 675 and the announcement of Find 688 might lead to the invention of a hovering vehicle? If I may be so direct, son, that's *my* job!" Barton said with a serious tone which startled LaDon until Barton's face slowly melted into a smile.

LaDon took a quick glance at the other Assembly members and noticed similar smiles on their faces. Aleen gave Barton a look of mock disappointment as if she knew Barton was simply up to no good. LaDon took the moment to shake the adrenaline rush and relaxed once more.

Blaine spoke up. "You're good, kid. That's why we need you."

"Yes, with this new technology, we need someone at the top of their game. Someone young, with new ideas...and old ones," Barton said as he stood from his seat. "You see, Mr. Grafter, the discovery of this new universe is currently perplexing us. We have yet to understand what we are seeing when the Solaspheres return from their destinations. There's no pattern to this variable that we can isolate. The variable, LaDon. Do you remember hearing talk about the variable during the meeting?"

Of course, LaDon had heard everything during that meeting. Every word still resonated in his mind as if he just heard them. The unknown variable and the Solaspheres jumping into another universe. All of these new ideas awaiting to be explored.

"Yes sir, Mr. Urthorn. I remember."

"This variable is the only pattern. If the variable is zero, the Solasphere records blackness. Every input sensor on the infernal device records nothing. Just cold, black nothing. If we move this variable into negative numbers, the Solasphere never leaves the planet. If we increase the variable into positive numbers, even the smallest decimal place, we see light rushing past the Solasphere so fast the sensor can barely register the visual. The higher the number, the more activity we can see and feel. After a while the light starts to fade and becomes spread out. Eventually we see stars similar to our own universe. As for sound, we only pick up faint waves of cosmic energy. As we set the variable higher and higher, the stars become dimmer and dimmer until there is total blackness. Placement of the device is not that difficult. It's this variable and the patterns of light traveling through space that we cannot grasp." Barton stopped to take a breath.

LaDon wondered if he should speak. He had nothing to say. His mind continued to race, trying to explain the variable.

"We want you to examine the Solasphere recordings. Watch them in sequence. Heck, you're the expert, watch them however you wish." Barton exclaimed as he took a few steps and propped back on the table just on the other side of his seat.

LaDon took this moment to drift into his

mind, taking everything he had just heard from Barton's lips. He tried to picture what the Solasphere recordings would show if lined up in sequence. *The first few Solaspheres show nothing at all. Next, brilliant bursts of light followed by an abyss full of stars. Finally, nothingness once again.*

LaDon looked up from his thought and asked, "Sir, just how many Solaspheres have been sent and how long do they record?"

Alex spoke up and quickly answered, "One hundred and twenty Solaspheres. Each of them recorded approximately fifteen seconds."

That's not enough. That's not enough devices and not enough recording length. LaDon imagined what the Solaspheres might see if they stayed longer in the new universe. He pulled on every bit of knowledge from his experience with the Solaspheres. He formulated different scenarios from the information he had been given up to this point. No matter what he tried, he could not seem to make a connection. *If I could just get a look at the recordings,* LaDon thought inwardly.

"We want you to start your work in the new lab immediately. This new lab is where the Solaspheres are being deployed to the new universe. We hope you don't mind, but we have arranged for your current employment to be temporarily discontinued. Don't worry, you've trained them well. In this new lab, we have installed a new system. One with your latest ideas in place. The setting should suit you nicely, Mr. Grafter. We hope you like it." Blaine stood, gesturing to LaDon that he was dismissed.

LaDon got the message. The meeting was over. He still felt the childlike essence inside him, thrilled to be in the presence of the Assembly. Now his adult nature brought on a completely different emotion. He felt a sense of duty. Even a sense of responsibility. He knew he would stop at nothing until he understood this mystery. Barton quickly took a few steps in LaDon's direction and placed his hand on his shoulder.

"We know you will do your best, LaDon. Is it acceptable for me to call you LaDon?" Barton asked respectfully.

"By all means, sir," LaDon answered as his heart beamed with optimism and merriment.

"And feel free to address me as Barton in a private setting such as this." Barton explained as they moved toward the door. "Everywhere else, I have to be Mr. Urthorn. I hope you understand."

Barton's hand dropped from LaDon's shoulder. LaDon realized the Assembly was remaining in the room, probably to discuss the meeting they just had with him.

"I've enhanced your Holopass to allow you access to the facilities you will need here at Nalkalin. Please remember you will need to reanimate your Holopass with your credentials as you travel in and out of Nalkalin's perimeter," Alex called from across the room.

LaDon's mind was still in a daydream state. He imagined Solaspheres zooming in and out of this universe. He pictured the sights explained by Barton just moments before. LaDon laid his hand on the door lever to exit the room.

Suddenly, a vision struck. It was almost as if LaDon's mind left his body for a moment. He felt like he was inside an actual Solasphere. He felt himself floating through the space of this neighboring universe, seeing the sights around him. At that moment, like a tumbler inside an old metal lock finding the ridges of that very special key, his thoughts clicked into place. He knew exactly what the variable represented. Once more, his pulse began to race. The door suddenly felt as heavy as a boulder. He stopped his attempt to pull the door open and turned slowly to face the Assembly. Their attention was on each other rather than LaDon. Obviously, they thought he had already left the room.

"Time." The word fell from his mouth almost as if he didn't speak it at all.

"What's that, dear?" Aleen said from a distance peering around Barton with an alarmed face.

"He said...time...", Barton answered, slowly turning to look over his shoulder at LaDon.

"The variable ma'am. It's time. I'm almost sure of it," LaDon said with even more conviction.

"Of....course....", Alex gasped as he squeezed his fists together like a fighter getting ready for a match.

LaDon watched as Barton looked away, propping himself up on the table with both arms. LaDon saw him glaring down at the table's surface. *Is he smiling? Is he upset?* Just as the thought crossed LaDon's mind, Barton turned slowly to face LaDon. Barton's face was not upset at all. It was unmistakably relief, almost as if Barton had hoped LaDon would figure it out before he left the room. LaDon watched as the smiling faces of the Assembly made their way around the table to join LaDon.

Barton looked at LaDon with an awestruck expression, "All the scientists on Solaya, stuck in their equations and variables. You didn't even make it out of the room, yet somehow you figured it out. Everything in my heart told me you were the right one to ask. I know two Solayans that will want to hear this news immediately."

Chapter 5

The Observadome

LaDon watched as Barton's focus slipped into his viewer. Barton swiped and poked a few times in the air to send a message with his viewer.

"How could you possibly have come to this conclusion? Scientists have been working on this for days." Alex protested as he paced back and forth.

"Maybe they were just looking at it from the wrong angle? Buried in the equations and science clouding their common sense." Blaine replied to Alex's question.

Barton's gaze came back to reality as he spoke, "I knew he would be able to come up with the answer. I tend to agree, Blaine. They've been looking at this all wrong. We simply needed another point of view. A point of view specifically from an observational perspective rather than a scientific one."

"My dear, what led you to such a grand conclusion? We have seen bursts of light and blackness. Why would the passage of time be your conclusion?" Aleen asked inquisitively.

Each Assembly member turned their undivided attention to LaDon, who was still trying to formulate his thoughts.

LaDon paused, looked at each member for a moment, and finally spoke, "It was the progression in which Barton explained. Blackness, bursts of light, scattered stars, and then nothing. Mix that with Alex's explanation of the volume of data recorded up to this point. I felt there was simply not enough data gathered. So I multiplied the number of data recordings in my head, assuming the recordings would continue this pattern. In my head, it began to resemble the formation of our own universe. I have a few friends in the deep space exploration area. I've often heard them discuss this particular matter. They match almost exactly, except for the blackness. I would be willing to guess if you adjust the positioning of the Solasphere a few times at a low time variable, such as one-tenth or even a single digit such as one or two, and perform a full spherical sweep of the view matrix, you will not see blackness. You would see one brightly burning point. This is said to be the point in which time begins for a universe. This is all theory, of course. I would need to see all the data collected up to this point to confirm any theories."

"And that is exactly what you will do," demanded Barton. "We shall continue everything as planned. We do not want to alert anyone to this conclusion just yet. You will report to the new observation lab. It has been fully functional for some time now, but the outer construction is just reaching its completion."

"Barton, should we call another public meeting?" Blaine asked.

"No, there is no need. We will, however, need to immediately call a meeting with the Lead

Representatives. Be wary; you know Joh Lin will want to know that a Vaknoreeyan played a special part in this discovery. We are never going to hear the end of it," Barton smiled as he ushered everyone toward the door.

"I'll get that meeting scheduled right away, sir," Alex responded as he moved toward his station.

LaDon had yet to grasp the full impact of what had just occurred. Barton was right. LaDon was now part of this discovery. The Vaknoreeyan people would beam with pride once they found out. LaDon's heart filled with a sense of worth. Even more than when he received the message from Barton in his viewer. As the remaining Assembly members ushered LaDon to the door, two fatigued-looking individuals, gasping for air, crashed into the room. Behind them two security guards sprinted frantically to catch up.

"Halt! Slow down there!" yelled a security guard.

"It's quite all right. No need for alarm," Barton's calming voice rang out over the noise.

"Sir!" one of the potential intruders said between pants. "Mr. Urthorn, your message. Is it true? Did someone figure it out?"

Barton smiled and placed his hand on the back of LaDon's shoulder and said, "Time."

The intruders abruptly caught their breath. Almost like someone had supplied them with new lungs. Their expressions went blank as if all knowledge they once possessed had now been removed from their minds.

"But how could you? I mean...you couldn't

possibly..." The other man finally spoke.

"My dear friends. Allow me to introduce someone very important to this project. This is LaDon Grafter, noted historian." Barton proudly introduced LaDon as he firmed up his grip on LaDon's shoulder. "LaDon, this is Jendall Kimnor and Phelix Castor."

LaDon immediately recognized the names while at the same time recognizing Jendall's face. LaDon's face lit up with a smile and he immediately reached out his hand to greet them. The two scientist's faces matched LaDon's smile as their puzzled looks found their missing pieces.

"AH, YES! The historian! Of course! Who better to solve such a riddle? Well, it is a pleasure to finally meet you. We were told you would be coming down to help us make some sense of the Solasphere recordings. The name's Jendall. I believe we've met," Jendall exclaimed as he greeted LaDon with a vigorous handshake.

"Yes, so nice that you remembered. It was quite a while ago. It will be a pleasure working with you both. I am looking forward to it. My most sincere congratulations on this amazing discovery by the way. I feel privileged to be able to tell you in person," LaDon sincerely replied.

"Pffft, you just figured out the variable! We should be congratulating you!" Jendall smiled as he reached out and patted LaDon on the side of the arm.

Catching the last of his breath, Phelix promptly followed Jendall's introduction, "Yes, indeed, it is good to finally meet you, sir. Sorry for barging in like that. We both got so excited when we saw Mr. Urthorn's message. We rushed up here as fast as we could. The name's Phelix Castor. We look forward to working with you."

"The pleasure is mine, Phelix," LaDon replied with a smile.

"I haven't even seen you in the lab yet," Phelix said with an inquiring tone. "That is amazing all by itself. Not one Solasphere recording and you still came to that conclusion. Astounding! Sometimes it's all about perspective, I suppose."

"Gentlemen, you couldn't have come at a better time. Please, let's step back into the room for a moment. I want to discuss this matter briefly before you return to the lab," Barton ordered as the group stepped just inside the door enough for it to close behind them. "I want this matter of the variable kept in the strictest of confidence. We are already setting up a meeting with the Lead Representatives tomorrow. Please keep this quiet until then. For now, Jendall and Phelix, if you would show LaDon to his terminal so he can begin his work, I would be greatly appreciative."

Everyone in the huddle agreed to Barton's conclusion and made their way to the exit.

As LaDon, Jendall, and Phelix left the room, Barton spoke from the doorway, "You are all a credit to your respective nations, my friends. Continue to make Solaya proud."

All three smiled and threw up their hands in appreciation to their respected leader.

Without a second thought, Jendall and Phelix accepted LaDon as a team member. They began to explain all of the work they had done. They left out any talk of the variable after the orders they just received from Barton. As they traveled to the new lab, a location within Nalkalin in which LaDon had yet to visit, even in a Solasphere, the two scientists filled LaDon's mind with the wonders of their newfound invention. He was able to follow some of it. He recognized the terminology from documenting the smaller scientific findings recorded from his Solaspheres.

As they entered the lift, LaDon noticed a new button on the control panel. A faint blue glow surrounded the button labeled OD. All of the other floors were still illuminated by a faint white glow. Of course, the new OD button was the one Jendall pressed. The lift responded instantly and whisked them toward the new lab. When the lift doors opened, they moved toward the only door visible once exiting the lift. The sign on the door read 'Observadome'. At least now he knew what OD meant. What could possibly be behind the door?

Although Jendall and Phelix's words had lost their meaning about half way through the lift ride, LaDon heard Phelix loud and clear as he opened the door of the Observadome and said, "You are going to love this."

Through the eyes of his Solaspheres, LaDon had seen many places across Solaya—the wondrous Prime Hall, the glorious Nalkalin, and even the beautiful landscape of the Grand Plaza across the sea. LaDon had thought there was nothing left to amaze him. He was wrong.

His eyes drifted downward from the dome-shaped ceiling to the rows of stadium seating surrounding the borders of the room. In the center of the room, Solasphere stations surrounded a cluster of equipment. The stations were just in reach of the terminals. This gave the user complete access to the Solaspheres.

Jendall and Phelix ushered LaDon forward, allowing him to go first up the small aisle toward the terminals.

"This terminal is setup for you," Jendall explained as he pulled out the seat.

LaDon slowly took his seat as if he was sitting on a very cold surface. He saw all of the controls he was accustomed to seeing. One thing stuck out over the rest. It looked much different than his usual workstation. A small set of controls to the left of his main panel labeled 'Placement Panel'. Jendall and Phelix both smiled at each other when they noticed LaDon staring at the new panel.

"This panel is where you input your desired coordinates of the Solasphere," Phelix explained as he leaned over the panel. "This lets you set the positioning. This one lets you set the duration of the trip. Finally, this one here...well this one is your variable. We didn't know what to label it until now. It's simple. You place the Solasphere on the pad, set the coordinates, the duration, and finally, the time."

"But that's not even the best part!" exclaimed Jendall as he connected a Solasphere that had just returned from the abyss. "These last two buttons on the new panel let you link your Solasphere to the viewing screens. This is for when you want everyone to see. Obviously you lose the sensory input function. See, watch."

Jendall pointed straight up toward the ceiling. LaDon smiled in absolute amazement for the first time since he arrived at Nalkalin. He had yet to even notice the ceiling after being blinded by the power of such a facility. All around him in a three-hundred-and-sixty degree view, LaDon saw multiple view screens in brilliant resolution. Some screens were joined to create larger screens. Some only displayed single recordings without stretching to other displays. The recording Jendall had connected for him took three displays. It showed a vast universe of beautiful star formations, unlike anything LaDon had ever seen.

"Here is where you can set how many displays you wish to populate." Phelix explained as he saw LaDon looking at all of the other displays.

LaDon began to fathom the possibilities of such technology at his fingertips. Ideas raced through his mind. *I can take one hundred Solasphere recordings and place them side by side on each screen circling the room. It would make for the perfect timeline. Then I could watch the formation of this galaxy unfold right before my eyes. Oh wow, the possibilities.* LaDon's mind was lost in the opportunities that awaited him. The room vibrated ever so slightly, and he looked up from his momentary reverie.

"What do you think about *this*?" Jendall said from one of the seats surrounding the room.

While LaDon gawked, Jendall had slipped off to a distant seat surrounding the Observadome. The chairs surrounding the room were rotated slowly around the room. This allowed someone to see all of the view screens from different angles. *Brilliant.* He itched to begin logging the Solaspheres and sending his first of many into the new universe.

"Well, I believe we've shown you all the buttons to press. Just give us a shout if you need anything. Our terminals are over here." Jendall jumped from the surrounding seating and over to his main terminal.

"Aren't you two going to show him what we've recorded so far? He needs to know where to start," said a voice from behind the farthest terminal.

LaDon hadn't even noticed anyone sitting at the far terminal. He stood from his seat to get a better view of the voice behind the terminal. He was glad he did. As the individual stood from their station, LaDon was shell shocked. Not by the wonderment of technology or the beauty of a vast landscape. This was beauty of a completely different nature.

"Oh yeah! Almost forgot you were here today, missy," Phelix exclaimed. The voice seemed to surprise him as much as it did LaDon. "LaDon, this is Larissa Sonne. She is Vaknoreeyan as well. She is here to assist from a deep space perspective. Since we seem to be sending the Solaspheres into deep space, they figured she could help. Larissa, this is

LaDon Grafter, our historian."

LaDon made his way across the room to respectfully greet Larissa.

"Ah, yes. The famous LaDon Grafter. It is a pleasure to finally meet you. I've followed your work closely. Those were some impressive improvements you made to the archives a while back. It got the Assembly's attention too, I see. So, you're the one who figured out the variable, huh?" Larissa asked with a cunning grin on her face.

LaDon was immediately confused. First of all, she knew him. Second, how could she know about the variable already? They had been careful not to say anything about the variable since they left the Unification Chamber.

"But how do you know...and how do you know that LaDon..." Jendall's words drifted as Larissa cut him off.

"Don't worry, boys. I can keep a secret. You were going to tell me anyway. I mean, I am part of the team. Besides, I've been sitting over here the entire time." Larissa stretched her words so LaDon, Phelix, and Jendall could catch up. "I believe you said, 'and this is *your* variable'."

All three let out a loud sigh as they recalled the comment.

"Well, I thought the room was empty. It should've been empty. You weren't supposed to be here today," Phelix protested, then realized he was out of debatable options.

"Sorry to disappoint. Besides, we're a team here. The information leak stops with me. Again, I was going to find out sooner or later, right?" Larissa smiled reassuringly at LaDon. "I figure they will make this information public anyway once the Lead Representatives are notified. Are they having another meeting?"

"Just a meeting with the representatives. Nothing like the meeting we just witnessed earlier today. And it's a pleasure to meet you as well, Ms. Sonne. You seem to know about me, which puts me at a disadvantage," LaDon explained as he felt himself becoming comfortable with Larissa's presence.

"Please call me Larissa. As for you two, aren't you going to show him what we've accomplished so far, or are you going to bore him with the story of your first Solasphere disappearing without a trace?" Larissa poked fun as Jendall and Phelix waved off her taunting and turned to LaDon.

"She's right. Let me show you what we have recorded so far and what our main objective is at this point," Phelix explained as he and Jendall made their way back toward LaDon's station.

"It is nice to meet you, Larissa. I look forward to working with you." LaDon said in his best professional manner.

"Likewise." Larissa shook LaDon's hand and quickly pulled him in close. "And don't let those two bother you too much. They can be a little eccentric at times."

LaDon met her eyes, smiled, and nodded. He turned and walked toward his terminal. He couldn't help but notice how utterly stunning she was. And she smelled amazing. He arrived to find Jendall and Phelix hovering over his terminal once again.

"All right, LaDon. Larissa is on the mark. We need to get you started. You need to see how far we've gotten. Now, from what I understand from the Assembly, this is the system layout you designed. We have been filing the recordings according to your specifications. Do you mind taking a look and tell us what you think?" Jendall asked as he leaned in to watch the master at work.

LaDon sat up and looked closely at the console. It looked just as he had imagined. He checked the filing system and naming conventions. They were all in perfect order, almost as if he had stored them himself.

"Gentlemen, I have to say, the filing and naming conventions are exactly to my specifications." LaDon smiled as he perused more screens, checking the settings of the application.

"Well, Larissa is the one that keeps the filing and naming conventions tidy. She's been known to throw things at us if we misfile something," Phelix muttered as he ducked his head and shifted his eyes over to Larissa.

"Stop being in such a rush and I wouldn't have to throw anything," Larissa shouted from behind her terminal.

She peeked over the top of her terminal and met LaDon's eyes. Phelix and Jendall babbled about the specifications, but LaDon already knew where to begin. This was good, because he was too busy staring to listen. He smiled at her, letting her know he could see through their tomfoolery. He could only see her eyes, but he could tell she smiled back.

"Guys, thank you very much for the tour and thank you for showing me to my station. I believe I've got it from here. I take it these are my Solaspheres?" LaDon held one up.

"Yes, yes they are," both scientists answered together.

They stepped back from LaDon as they seemed to notice they were crowding him.

"Larissa, can you hear me?" LaDon directed his voice toward her station.

"Yes, I can." Larissa answered in return.

"All right, I would like to rearrange these recordings based on the time variable. Now that we understand the variable, this gives us a time frame. I would like to order these recordings by that field. I want to start recording from zero with the time variable and increase from there. I will start with zero to five. Someone else take five to ten and so on and so forth. I want to go through at least two or three Solaspheres before we decide how many decimal places we need to take variable. I want to know if setting the variable to a whole number correlates to one second, one minute, one month, one year, etc. Understood?" LaDon explained with confidence.

LaDon was in his element. He had yet to show any assertiveness until then. He quickly scanned the room to check the faces of the team. He was always careful not to overstep his bounds.

"I hope this acceptable to everyone. I don't

want to impede current progress or cause hurt feelings, but with the new information of the variable, I know this will be a more effective method of approach. It seems the recordings are currently incrementing by the positioning variable, am I right?" LaDon asked.

All heads nodded up and down in unison.

"No complaints here. After all, you're the boss," Larissa said as LaDon's face contorted into confusion.

"Boss?" LaDon cocked his head to one side in disbelief.

"Unless someone else is controlling my viewer messages," Larissa glanced out of reality to check her messages again.

LaDon, Phelix, and Jendall each checked theirs as well. They too had just received a message from the Assembly explaining this news. the Assembly had placed LaDon in charge of filing, storing, and archiving all data.

"Are you all okay with this change?" LaDon asked as they all returned from their viewers.

All three, once again, nodded with enthusiasm as if they were hoping this was the case.

"This is your ship, my friend. We have the privilege of getting to use this new facility and actually meet the creator of this storage software. Now we get to work for him? We were hoping you were the one to take the lead." Phelix explained. "Letting me run this project would be like letting a barber repair your transport."

The entire room erupted in harmonious

laughter as LaDon let this sink in. He was the point person. Inside, LaDon knew he was ready for the challenge. More so, he was anxious to venture into a Solasphere recording and see this new plane of existence. LaDon looked within himself for the knowledge and strength to do his best. Not only for himself, not only for Barton Urthorn and the other Assembly members, but his grandfather too. *If only he could see me now.*

Chapter 6

Boys and Their Toys

"All right. We have the time variable as a point of reference, but this makes the recordings up to this point seem sporadic. Probably because you guys were using the placement variable as the sequence of events. Now we can watch each recording in time sequence. We can lay the recordings side by side, giving us a continuous video feed of the creation of the universe. As for the existing recordings you've made up to this point, we will keep them. As we reach their point in the timeline, we will celebrate in the fact that we can skip that particular recording."

LaDon paused for effect as the room gave a small chuckle. "I am going to send a few Solaspheres in at a low time variable. I need to know whether the blackness is before the initial expansion of all matter in space. If I am right, there is not just blackness at zero. There should be a brilliant light source somewhere in the universe. We just need to adjust the positioning variable a few times to find the location of this source before it expands. Does this make sense so far?"

The room agreed as LaDon reached for a Solasphere. He noticed these Solaspheres were sturdier than the ones used for planetary surveillance. *Obviously these must be reinforced to handle the punishing environment of deep space.*

He connected the device to his station, adjusted his seat, and configured his terminal for his first Solasphere recording. He examined each variable closely and set each one accordingly. First, he noticed the preset recording length. It defaulted to fifteen seconds. Next, he adjusted for time. He set the variable to one one-hundredth. Finally, he set the positioning variable to a different coordinate than one used previously. He looked up at his team. He wondered if they could see the anticipation on his face. He reached toward the panel and executed his settings. The machinery whirred into a frenzy, and then the Solasphere twitched. A thin glass shield arose from the table and engulfed the device. A dim, blue light encased the device as the energy level increased. LaDon looked to the team for assurance. This was quite different than his normal Solasphere encounters. He needed reassurance that everything was going as intended. He saw faces of patience and calm. He turned his attention back to the Solasphere. Just as his eyes found it, it was gone. At that moment, for LaDon, time seemed to slow. He couldn't believe what he just witnessed.

"Remarkable!" LaDon exclaimed.

"Thank you," answered Phelix and Jendall in unison as they smiled at each other with pride.

Before LaDon could sit back in his chair and take in what had just happened, he heard another sound. It was similar to the sound he had heard moments before but more distant. A faint glow appeared where the Solasphere had just vanished. LaDon watched the enclosed pad. A dim, glowing nimbus about the size of a Solasphere began to form inside the enclosure. In a brilliant flash, the device materialized before his eyes. LaDon had never seen anything like this. He was astonished. Just days before, he would have considered this magic. Now, seeing it with his own eyes, the findings report from the Assembly did not do it justice. LaDon looked to the team for the okay before reaching for the Solasphere.

"You have to give it just a few minutes to warm up. The environmental settings inside the casing bring the Solasphere back to acceptable temperature levels. They return from the clutches of space at extremely cold temperatures. The casing will not drop until all readings are within safety parameters. You can go ahead now, it's safe," Phelix urged.

LaDon took the Solasphere in his hands and examined it for any marks or blemishes. It looked exactly like it did the moment it left the pad. Like a child with a new toy at Christmas time, LaDon spun around in his seat and attached the Solasphere to his terminal. He logged everything appropriately and started the transfer of information. Finally, he activated his terminal to view the recording.

"By default, we disengage the sensors allowing for touch so we do not experience the cold. There are plenty of safety measures either way, but we like to be safe," Jendall explained as he watched with enthusiasm. "Hurry, we want to watch you experience it for the first time. We know you are familiar with Solaspheres, but it's amazing to see.

But you will probably see blackness this time. Either way, give it a go!"

LaDon adjusted himself in his seat to view his first inter-universal viewing through the eyes of a Solasphere. His terminal displayed the list of recordings. He activated the only file on record at this terminal. He closed his eyes and relaxed in his chair as he had done many times before. He felt his body move away from his seat. LaDon knew he was still seated at his new station, but this was the feeling he was used to when he first started a Solasphere recording. His heads-up display immediately came into view the moment the recording started. At the top right of his display, he could see the seconds ticking by. Almost instantly as the recording started, LaDon began looking in all directions. He saw nothing but blackness. Blackness all around him with no light in sight. The timer counted down, thirteen seconds, fourteen seconds, then a brilliant flash. The recording had ended. LaDon mentally disengaged himself from the recording and sat up in his seat. He looked around the room with a perplexed expression as he stood from his seat.

"Don't worry. It's black for us too. If you increase the time variable, you will start to see stars. Do you want to watch one of my recordings?" Jendall asked.

LaDon didn't respond. He slowly walked away from his terminal and over toward one of the seating areas around the room. Running his fingers through his hair, LaDon leaned over one of the chairs and tightly closed his eyes. He began mumbling to himself. *Blackness, blackness, but why?*

"That's what we want to kn..." Jendall began, but Larissa jarred him on one side.

"Shh, he's thinking. Give him a moment. Imagine how you first felt when you saw one of these recordings," Larissa whispered to Jendall. He lowered his head in submission, as if suddenly aware of his loud mouth.

She was right. LaDon was already getting good at drowning out Jendall's excessive talking. He was thinking hard about the blackness. *Something does not come from nothing. There has to be an answer,* LaDon said to his inner monologue as he continued. *It can't be all black. There's something there. Why can't we see anything?*

LaDon began thinking about a time when he and his grandfather would lie under the stars and watch the Solayan sky. He remembered lying there when there was still a little light left in the sky. Slowly, more and more stars would come into view.

"Look, I can already see some stars, grandfather! There's one. There's another!" LaDon could hear his younger self pointing up into the heavens.

"I see them too! But just wait, LaDon. More and more will start to appear." Pomph's voice echoed through LaDon's thoughts.

"Why is that? Why can't I see them now?" young LaDon had asked.

LaDon's mind sprang from its daydream and he walked briskly toward his terminal.

"That's it! Yes, of course," LaDon exclaimed.

"What? What's it?" Phelix asked, seeming startled by LaDon's spark of passion.

"The recordings. They need to be longer. A full one minute should do it. Is this possible?" LaDon blurted as he reached for another Solasphere.

"Well, um, of course, I suppose. We set the recording length to fifteen seconds for safety's sake. We felt it risky to leave the Solaspheres away for much longer due to the chances of them not returning. It just kind of stuck from there." Phelix explained as he assisted LaDon with connecting the Solasphere to his terminal.

"I am sending this one out for one minute. If I understand the Observadome's capabilities, I can broadcast this recording to every display in this room. This would give me a full three-hundred-and-sixty degree view of the recording, correct?" LaDon asked looking toward the team.

"Ha! I've got something even better than that." Jendall said with a smirk as he reached over to his station and pulled out a small remote. "Be amazed, my friend, be amazed."

The room began to vibrate under LaDon's feet. The seats surrounding the Observadome folded and dropped into the floor. The walls shed their skin and the floor opened up to reveal a transparent viewing area. The entire room was now a spherical viewing facility. LaDon saw this and quickly realized he could watch the recording as if he were inside the Solasphere looking outward.

"Those Nuheeyans just keep getting better and

better." LaDon smiled as he admired the beauty and technology surrounding him. "Okay, I am sending this one to the same location as the one I just sent, but I am sending it for one minute."

Larissa stepped forward. "Not to spoil any surprises LaDon, but can I ask what you plan to see that you didn't see before?"

"The Solasphere's viewing sensor. When it leaves this plane of existence, the first thing it sees is a brilliant light show. This obviously comes from being teleported. Since these devices operate much like our own eyes, in order to be compatible with our brains, they need time to adjust to a new light setting. If we let it adjust, it will slowly be able to take in light from greater distances. It needs more than fifteen seconds to do this effectively." LaDon explained as if he was teaching a class.

"That makes perfect sense! The longer you stare into the blackness, the more your eyes will adjust to the darkness. This would give the time needed to pick up any shred of light struggling to come through. All right, it's ready LaDon," said Phelix as he finished connecting the Solasphere.

LaDon turned to his terminal and configured the controls. He brought up the previous configuration, altered the length of time the device would remain, and activated the program. Within a matter of seconds, the Solasphere evaporated. Just a few seconds later, it reappeared.

"Wait, I'm confused." LaDon looked at his team in bewilderment. "It was supposed to be gone for a full minute, yet it's back already."

"Yes, it is programmed to exist in the new universe for the time allotted, but it will return to this world within just a few seconds after it left," Jendall answered. "The math in our equations doesn't allow for a device to return before the time it left. If it were to try and return to our past, it would vanish. I believe it has something to do with a time paradox, but I'm not the expert in time travel. I just work here."

LaDon had no choice but to believe him.

"I don't feel I am done being amazed today," LaDon smiled as he reached for the Solasphere.

LaDon uploaded the recording to his terminal. The download completed instantly. LaDon admired the new archiving system. After all, he had aided in the design of the new processing capabilities.

Phelix reached in to show LaDon the key combination to display the recording to the entire dome. The lights dimmed, and they were all surrounded by blackness.

"Now, if I am right, somewhere toward the end of this recording, we should begin to see something," LaDon said as he turned himself in all directions.

The team spread out to observe the entire room.

"Thirty second mark," Jendall announced as the timer ticked away. "Forty second mark. Fifty second m...."

"*There*!" shouted LaDon.

The team rushed to see where he was pointing. Before they could make it to his side, the recording went black.

"I'll play it back," Larissa called as she hurried to LaDon's station. "There, it's playing. I'm moving the recording forward to the forty-five second mark. Starting the recording."

"THERE!" LaDon shouted again as if he was seeing it for the first time.

"Why is that the only one?" Phelix wondered aloud.

"I have some friends in deep space that discuss the creation of our universe. They say it starts from a huge explosion in space occurring from one single point." LaDon explained as if he is trying to believe it himself. "Let's adjust the positioning of the Solasphere. Same time variable, but let's see if we position it closer to this star."

"I'm way ahead of you LaDon," Larissa peered from over LaDon's terminal with a sly grin. "The deep space girl is right in sync with you on this one. The positioning variable has been tested a few times. We believe we understand it quite well. I have a Solasphere ready. This should at least move us in the right direction."

Larissa's fingers moved over the console with ease. The Solasphere shimmered and poofed, and then returned with the same flash of light.

"I'm connecting it now guys," Larissa explained as the moved the Solasphere into place for uploading. "Starting playback."

This time the light source was much larger than before. Huge, in fact. LaDon glanced at the timer in the top right corner and noticed ten seconds. Suddenly the recording went blank. The entire room knew this was not a full minute of recording time.

"What happened? Larissa, did you stop the playback or did you change the amount of time it would stay over there?" LaDon looked toward his terminal for a response.

"No, I didn't touch anything. Same settings as before. Same time and the same length of stay. Only a different location." Larissa looked at the station as if she had never seen the controls before in her life.

"Check the sensors. Did anything overload?" Phelix hurried to the terminal to view the configuration readings of the Solasphere. "Oh my. The safety feature activated and sent it back early. It was heating up extremely fast."

"A heat source? The only thing we could see was the light," said Jendall as silence overwhelmed the room for a moment while the team pondered the situation.

"The light source must be emitting massive amounts of heat. That must be it—the single point from which the universe starts. It is said to consist of all matter and energy in the universe. Just being that close, the Solasphere almost went up in smoke," LaDon gasped as the realization hit him. "If I remember correctly from the meeting, the finding report said you guys have had Solaspheres that never return, right?"

"This is true." answered Jendall still a bit perplexed.

"I get it. Watch this." LaDon connected a different Solasphere.

The team leaned over LaDon's shoulder as he configured the terminal. With a few clicks of the console, LaDon sent the Solasphere away once more.

"You're setting the Solasphere to the same location, but upping the time variable to one-tenth?" Jendall asked curiously.

"We'll never see that Solasphere again," explained Larissa as her eyes met LaDon's and their minds connected.

"Of course! You're trying to put it in the path of the explosion just as all the energy rushes out into space,", Jendall said in a high pitch tone as he spun in a circle like a child learning a new fact of life. "But why one-tenth?"

"Trying my luck. I don't really know when the blast occurs, but we do know that time starts at zero since inputting a negative time variable causes the Solasphere to go nowhere. This is simply because there is nowhere to go." LaDon explained as if placing in the last piece of a jigsaw puzzle. "We just need to send out a couple dozen Solaspheres with the right configuration to judge how the time variable measures up against how fast time is traveling. I will take this task upon myself. The rest of you will continue as planned. Stay clear of the lower numbers for now. Although we have plenty of Solaspheres, we don't want to send hundreds to their death. I will map out the dangers of recording so close to the explosion while you all start at a reasonable time variable and work your way up from there. Split the time amongst yourselves as you see fit."

The team nodded and moved to their stations

to continue their work. LaDon stopped to ask a question.

"I have yet to read my message from the Assembly. What I am about to ask may be there. How often do you all report to the Assembly?" LaDon asked.

"They asked us to report daily before we return home, but to report any significant findings immediately. I see where you're going with this. In my opinion, this news qualifies as significant." Phelix explained.

LaDon and the rest of the team realized they probably needed to report this at once.

"I agree with Phelix. Should we go now?" Larissa asked with concern.

"Yes, I believe so," LaDon agreed.

They each rose from their stations and LaDon activated the communication section of his viewer. "Mr. Urthorn, are you available?"

After a few seconds, LaDon's viewer illuminated.

"Ah! Mr. Grafter. Yes, I am. What can I do for you?" Barton answered with a delighted tone.

"Sir, we have some news that we feel is worth of the Assembly," LaDon recited into the air respectfully.

"Why not just run up here like crazy, mad scientists and tell us? You could call it a pattern after the second time," Barton asked jovially with a smile in his words.

LaDon looked toward Phelix and Jendall with a smile. He remembered them crashing into the room with the big news. Barton was right. They did appear like mad scientists. Phelix and Jendall had no idea why LaDon was smiling, so they both awkwardly returned the smile.

"I'll remember that for next time, sir. You can count on it. Is the Assembly available?" LaDon asked politely.

"Yes, we are. We are just finishing up preparing for tomorrow's meeting with the representatives. We were about to wrap up. It's been a long day. I know all of you are probably as tired we are. We will let this be your daily report. Also, I suggest we all call it a day after that. Please, come, report your news at once," Barton said as LaDon's viewer went black.

LaDon looked toward the team and said, "Well, let's go see the Assembly, shall we?"

LaDon was still getting used to the fact that he was now in direct communication with the very person that he admired more than anyone living on Solaya. The team fed off of each other's anxiety as they prepared to report the news to the Assembly. Phelix came up behind LaDon and whispered in his ear.

"He commented about us storming the meeting earlier and making a fool of ourselves, didn't he?" Phelix muttered.

"Yes, yes he did, but all in good fun," LaDon smiled as he walked along beside Phelix.

"Thank goodness he has a sense of humor. I bet Blaine thinks I'm an imbecile." Phelix lowered his head in shame.

"Eh, you were excited. It was ground breaking news. You were doing your job." LaDon attempted to reassure Phelix.

LaDon could see his words made Phelix feel somewhat better.

Down the lift they went toward the Unification Chamber. The huge double doors, once very heavy to a frightened LaDon, were now light in comparison. He felt as if he belonged. He had a sense of purpose. He had a real reason for being in this meeting room. As the team entered the room, the Assembly members greeted them warmly. There were exactly four chairs already setup in front of the Assembly's table awaiting their arrival. They each took a seat as they prepared to deliver their report.

Chapter 7

A Long Day's Work

"Ah! Welcome. Please have a seat," Blaine announced with a bright smile. "Can we get you anything? Food or drink? There's water here on the table if you'd like."

Jendall and Phelix both poured themselves a glass. After he fetched his own, Phelix offered to prepare a glass for LaDon and Larissa. They both politely declined. LaDon could not bear the thought of anything on his stomach during a meeting with the Assembly. He was anxious to relay the news they had just learned. To be able to document the initial creation of another universe could lead to the understanding of their own universe. Possibly even their own existence. There was really no end to the possibilities.

"Very well. Let's begin. I know all of us are ready to retire for the evening. So, let's look for what little energy we have left and make sure this news is relayed appropriately." Blaine let out a deep sigh and shifted his position to achieve more comfort from his seat. "Who would like to start?"

Without hesitation Jendall spoke first, "I believe LaDon should take this one, sir."

"Well we all..." LaDon began.

"This is not the forum for modesty, LaDon," Larissa cut in. "Speak."

"She means it. Otherwise, you are in danger of those flying objects I mentioned earlier. Plus, with this sitting arrangement, I am in the line of fire," Jendall stage whispered loud enough for everyone to hear.

LaDon worried that such behavior might be inappropriate in front of the Assembly. But everyone smiled, so his worry dissipated quickly. They were obviously accustomed to Jendall's behavior.

"Mr. Grafter, it seems you have been elected spokesperson. Congratulations. Now, out with it so I can go in search of my bed," Blaine said as his smile slipped away from his face and a weary expression took its place.

"Very well. First of all, I would like to say the Observadome is quite impressive. All of the controls are just as my design specified. Also, the room itself, the viewing area, is breathtaking. I believe the Nuweeyan people will never run out of tricks," LaDon rambled on as the room nodded in approval. "After familiarizing myself with the controls and observing the wonders of the establishment, my main focus was to understand the blackness when setting the time variable equal to zero. After sending one or two Solaspheres, again recording nothing but blackness, I remembered something my grandfather taught me. To save us a trip into my past, I'll cut this short. He would remind me that I could not see the stars until dark. He would go on to explain how more stars would slowly appear as the daylight turned to darkness."

"Well, yes, of course. It takes time for our eyes

to adjust to the darkness," Barton agreed eagerly.

"Exactly. And knowing that the Solasphere's optical sensor behaves similar to our own eyes, I deduced that the Solaspheres did not have enough time for the optical sensor to adjust to the blackness. We sent another unit, this time leaving it in the other universe for a full minute. Approximately fifty seconds into the recording, we saw it." LaDon ended his sentence with a snap of exuberance and a smile.

LaDon realized with pride that he was telling this story like his grandfather would tell it. LaDon also felt his team on each side of him reliving the moment as his story progressed. the Assembly did not speak. They were waiting for him to continue.

"It was the single point where all matter originates for that particular universe. Just as our deep space friends here on Solaya have discussed, the formation of this new universe came from one center point of energy and matter. We are now working on adjusting the time variable to understand the rate in which time passes. Without the technical speak, we want to know how much time passes in this new universe if we set the variable to one, two, three, and so on. Knowing this will teach us much about the passage of time and this universe's life span." LaDon felt like he'd just given an entire lecture at the lecture hall.

"Life span of the universe?" Alex questioned as curiosity crept onto his face.

"Yes. It is believed that the universe begins from a single point. All the matter and energy in the universe explodes out into space, and, over time, the matter and energy spread into space. After a while, all of the energy would eventually burn up and you are left with hunks of floating matter. Without energy, all of the stars would cease to shine and there would be no more light. This would mark the end of that universe. That was why the Solaspheres sent with extremely high time variables had recorded nothing but blackness as well. It all adds up," LaDon said with a degree of certainty in his voice.

"Would you be willing to bring this information before our deep space team? We would like to get their take on this. Also, I bet they would be thrilled to watch these recordings," Barton said in an awed voice.

"That would be great, Mr. Urthorn! They will definitely want to see the recordings. It would also give me time to catch up with old friends," Larissa exclaimed with an excited smile.

Aleen pleasantly responded to Larissa's excitement. "That's right, Larissa. You are from the deep space area yourself."

Larissa nodded as Blaine reclaimed the conversation. He was always prepared to move to the next matter at hand.

"Well, that settles it then. We will discuss everything we have learned up to this point with the Lead Representatives tomorrow. Until then, you are all aware this information is strictly confidential until deemed otherwise," Blaine quickly explained, once again showing his eagerness for the meeting to end.

Barton smiled at the team proudly. "I know each of you have been working hard since this new finding occurred. You deserve some rest. Don't make me have to chase you out of that dome. As for you, LaDon, I know you have just tasted this new adventure and are probably eager to map the history of this new plane of existence, but I'm saying this to you too. Go home!"

"If you all have nothing more to add to today's report, you're dismissed," Blaine said and rose from his seat.

Everyone stood from their seats and started their journey toward the door. One by one, they exited. In the end, the last two to leave the room were LaDon and Barton. As they journeyed down the hallway, Barton stepped forward to walk alongside LaDon. The two of them met eyes for a moment. LaDon felt that familiar, awestruck feeling of being in the presence of the mighty Barton Urthorn. This time it faded almost as quickly as it started.

Barton, who always seemed at ease, said, "You've impressed me today, LaDon. You have fell into this role quite easily. The team already seems to trust you. Of course, you are doing an excellent job with the research and progressing the technology. I had no doubt about that. But getting such a diverse set of individuals to trust you so quickly shows character. Confirms I picked the right Solayan for the job. Besides, this is always one of the greatest challenges we face as an assembly. Picking the right person for the job is a gamble in most cases. You have had little dealings with these types of surroundings, without being behind the eyes of a Solasphere of course."

"Well that's very kind of you to say. I suppose it comes from an old saying I used to hear from someone very dear to me," LaDon began to explain.

"Your grandfather, I presume?" Barton cut in, nodding.

"Why, yes. How did you...?" LaDon stopped walking and gaped.

Barton stopped as well.

"I am familiar with each of your profiles. They are fresh on my mind, so I assumed. Not to mention, I used to work closely with your grandfather. I'm sure he told you at some point," Barton answered as they continued walking.

"I do remember him saying that. He never went into much detail. I asked many times, because I loved hearing stories about the Assembly. He would always steer the conversation in a different direction or tell me it wasn't important," LaDon. Recalling it, he wondered why his grandfather had never expounded on his dealings with the Assembly.

"He was a great man, LaDon, but I don't have to tell you that. You remind me a lot of him. Same work ethic, same mannerisms, and even your terminology. At times, I think I can hear his dialect in your words. But, I interrupted. You were about to tell me something he used to say," Barton apologized.

"No, no problem at all. He used to explain to me how being historians might cause us to look only to our past for answers. He would follow by saying that we must focus on the present just as much as the past, because what we do here, in the present, is just as noteworthy. He meant this as a way to teach me to be mindful of those around me and not to ignore their needs. Also to focus on the present situation. 'To live in the moment,' if I recall his words correctly. This was not a simple lesson like some. This one took some time to sink in." LaDon recited this back in his head to make sure he had explained this lesson appropriately.

"That's a very valuable lesson. One that not all Solayans are lucky enough to have been taught," Barton huffed a laugh from deep within his gut. "Well, nevertheless, it is good to have you on our team. A Grafter has never failed me yet." Barton smiled and winked at LaDon as they reached the lift.

LaDon stepped on first, then Barton. They made small talk as the lift took them to the first floor. As they exited, LaDon realized Barton was headed in the opposite direction, toward the living quarters in Nalkalin. Barton and LaDon exchanged words as they parted.

"Have a good rest sir," LaDon said.

"And you as well, LaDon," Barton kindly replied as he began to walk away.

"Sir?" LaDon spoke quietly, but his voice slightly echoed through the vast, empty lobby.

"Yes?" Barton answered softly as he turned toward LaDon.

"If this is not an imposition, I would like to ask if one day you could tell me more about how well you knew my grandfather," LaDon asked, hoping for any bit of information he could get.

"When the time presents itself, of course. I'll tell you all there is to know," Barton responded in a fatherly tone.

They exchanged a smile, and with a gentle nod of LaDon's head, they went their separate ways.

This was a familiar feeling for LaDon. His grandfather, he remembered, would always dance around the question. For some reason, LaDon was never persistent. He thought for a moment to turn around, call Barton back, and insist he tell him right then and there. But that would have been rude.

He had only been on personal speaking terms with him for a single day, but felt he could trust him with his most valued possession. His memories. Just as he opened the exit door, in the distance, he heard the door to the Assembly's living quarters click open and shut.

LaDon realized it wasn't that late. Although he had taken in more information in one day than in recent years, he fought the urge to swing by the Observadome for one more Solasphere run. After replaying Barton's command to go home in his head, he came to the conclusion that he had plenty of time for that tomorrow.

LaDon looked into his viewer for the time. He had a little time to kill before heading home. The other team members were already out of sight. LaDon assumed they all went home. He began his journey back toward his transport. Remembering he had a long walk ahead made him happy. It gave him more time to think. His journey led him past the Prime Hall and eventually back to his office at the archives.

The trip back did not seem as long as his

original trip to Nalkalin earlier in the day. With his mind full of everything he had seen, time passed much more quickly than he was accustomed. As he reached his transport, he decided he would take the long route home. He debated on stopping at a quiet tavern to have a drink and think about the day's events. The low hum of his transport filled his ears as he pulled out on the main road. The two moons of Solaya shone brighter than most nights. He hadn't noticed the stars lately, but his new job was causing him to look toward the sky once again.

His mind drifted back to the blackness in the Solasphere recordings. The sound of silence when he was immersed in the recording. Pure silence was one of the loudest things he believed he had ever heard. One sensory input he hadn't given much thought to was taste. His taste buds simply felt as if they were not working during the recording. When he thought it through, there was nothing to taste. No gases, no liquids, nothing. It was a unique experience to say the least. He'd never experienced anything remotely similar to being in outer space. Feeling the weightlessness was a surreal experience. He was so focused on what he was seeing, he had paid no attention to those sensations for the time being. He told himself that tomorrow he would pay more attention as he mapped the birth of the new universe. Just the thought of watching the brilliant colors and spectacular light show of such an event filled him with even more excitement.

As he reached his town, he saw the small tavern he'd been debating about earlier. He thought twice and drove past the tavern toward home. He'd been around people enough today and deserved a little time away from civilization. Just as he made this decision, his viewer illuminated with a transmission from an unknown person. He tapped the air to answer the request.

"Yes?" LaDon spoke as the transmission synced.

"LaDon! It's me, Larissa. I wanted talk with you after we left the meeting, but I saw you talking with Barton. I didn't wish to disturb you," Larissa began with a very chipper tone.

"Ah, it would have been no problem," LaDon lied, as his conversation with Barton was personal in nature.

"Yeah, uh hunh. You guys looked pretty serious. Anyways, listen, while you were talking, Jendall, Phelix, and I talked about doing something together tomorrow night after we got done playing with the Solaspheres. We thought about it tonight, but we weren't sure how long you and Barton would be, so we decided to go home. So, how about it? Tomorrow night?" Larissa asked excitedly.

"I would love to. That sounds great," LaDon replied.

I do need to get out more, LaDon thought to himself.

"Great, I will let the two goofs know. We will plan tomorrow where to go. That way we can get everyone's input at one time," Larissa said with an excited chuckle.

She seemed to enjoy making plans.

"Very well then," LaDon said trying to end the conversation.

"You're not disconnecting that quickly, mister. So, tell me, did you enjoy today? Jendall and Phelix weren't too annoying were they? Don't worry, you just have to get used to them. They mean well." Larissa chatted away as if LaDon was a long lost friend.

LaDon picked up on the fact that Larissa was interested in continuing their conversation. This gave him a small boost of confidence as he searched for something witty to say.

"Today was extraordinary to say the least. To change jobs, unannounced, in one day, was a little overwhelming. And then the experiences in the Observadome? I'm on systems overload," LaDon explained.

He cringed after hearing his terminology lean toward geek. *Maybe she'll let that one slide. After all, she IS a scientist. Focus, LaDon, focus.*

"Tell me about it. My first day was all a jumble. They brought me in ten days ago. They showed me this wonderful invention, explained to me the situation, and then told me not to tell anyone. Keeping something like this to myself was probably one of the toughest things I've done in a while. It's nice to finally have someone else to talk to about it. Especially someone that has experienced it with me. Oh no, am I talking too much?" Larissa paused as she waited for LaDon to respond.

"Ha, not at all. I want to hear what you have to say," LaDon reassured her.

"Oh, well good. I have been told I talk a lot when I am excited about something. I suppose you are on your way home?"

"Yes, I'm just about to arrive," LaDon responded.

"All right. I will see you bright and early tomorrow. I'm looking forward to tomorrow night. Don't back out on us, now," Larissa said with a smile in her voice.

"Oh, I'll be there. Count on it. Have a good night, Larissa."

"You too."

The transmission went silent.

As LaDon exited his transport, he found himself attracted to Larissa. She was easy to relate to and definitely easy on the eyes.

With that in mind, LaDon entered his living quarters. Everything was like he left it earlier in the day. When he departed for work that morning, little did he know he would be working for the Assembly when he returned. The room almost felt as if someone else had lived there before, and now a new Solayan was occupying the residence. Things were different, very different. He had the Assembly to thank for that.

As he prepared for bed, he pulled back the sheets and slid between the covers. He was restless at first. He thought about the team, the Solaspheres, and where he would start the recordings in the morning. As he finally drifted off to sleep, his last thoughts were of his conversation with Barton. He felt a sense of protection from Barton, a feeling that he hadn't felt in quite some time. Was it a feeling of respect? Maybe a feeling of admiration? Whatever it was, he knew he had nothing to worry about when it involved Barton Urthorn.

Chapter 8

One Thing at a Time, Please

Early the next morning, LaDon finished his breakfast and headed toward his transport. His mind raced with events from the day before. In the span of a day, he had taken in a plethora of information. Not to mention the fact that he didn't even get a good chance to say farewell to his colleagues back at the archives. For this reason, LaDon woke up a little earlier than usual.

On the way to his old office, he paid special attention to the scenery. It would be the last time he took that route to work. So many years at the historical center, so many Solaspheres, and so many recordings. He knew his team would be sad to see him go. He had trained most of them since their very first day. There were plenty of worthy historians that would be able to fill his role. They would be fine, just as Barton had expressed the day before.

With old memories and new ones running through his mind, he arrived at the historical center. Not many Solayans had made it to the office just yet. He was a bit early, although there were a few early birds attempting to catch their worms. LaDon felt they were overdoing it. Either way, to each his own.

He exited his transport and started his trek toward his old office. Taking this final trip, everything seemed different and even a bit smaller. When compared to the wonders of Nalkalin, his old office grounds seemed plain. No elaborate structures. No beautiful scenery. Just the familiar square building and the same old fountain where he spent many lunch hours taking in the fresh air. He mainly wanted to pick up his belongings. His footsteps echoed a little louder than usual since the parking structure was a little more barren than he was accustomed. He accredited this to his early arrival.

As LaDon entered the building, he walked down the familiar hallway. He couldn't help but continue thinking how this would be the last time he traveled these halls. This would be the last time he entered his office. He rounded the corner to see his office door slightly ajar. He stepped up his pace to investigate. LaDon peered around the entry way of the door. His office was barren. It had been stripped clean of all equipment, furniture, and even personal belongings. Feeling a bit violated, LaDon looked around quickly outside his office. He noticed a light on a few doors down in the office of a co-worker, Jain Brelin.

"Jain, are you there?" LaDon spoke loud enough for her to hear him.

"Yes, I am. LaDon, is that you?" Jain replied excitedly.

"Yes, it is. Don't get up. I'm coming to you." LaDon hurried down the hall toward Jain's office.

"Hi LaDon! I didn't know if we would see you

today. We were told you were called away on important business with the Assembly. Is that true? I am so excited for you!" Jain approached LaDon and hugged him fiercely.

"Well, uh, thank you, Jain. That's most kind. And yes, this is true. One question. My office? My things? Everything!" LaDon asked, alarmed and puzzled.

"Oh, yes. Some of the Assembly's staff workers came yesterday while everyone was at the meeting. I was still here finishing up some paperwork. I figured I could read the report later and hear plenty of the gossip once the meeting was over. I saw them cleaning everything out and asked them what was going on. They said you had urgent orders to report to the Assembly and that you were moving locations. I figured you knew," Jain exclaimed, matching LaDon's puzzled expression.

"Well, yes, I did. I mean, I found out during the meeting. Wow, they move quickly, don't they? What if I had said I didn't want to take the assignment? Pfft, what I am thinking, of course I would, right? But could they know that? I wonder if I even had a choice in the matter." LaDon wondered aloud as Jain simply gathered gossip for another venue. "Well, I guess there's no need to stay here. I'll miss you, Jain. I hoped I would get to see more people so I could say farewell."

"Oh, you know you're welcome around here anytime. Just stop by when you are not off doing your super-secret Assembly work," Jain said in a comical tone still showing her excitement for LaDon.

"I most definitely will, Jain. It has been a pleasure working with everyone here." LaDon took a long look around and let out a small sigh. "It's just so sudden, you know?"

"Yes, it is, but if anyone can handle this, I know you can. You have taught us all so much. We knew this day would come. You've always been destined for greater things. Someone finally saw it. And for it to be the Assembly, well that makes it that much better! We will miss you, LaDon. Please come see us whenever you can," Jain said as she finally slowed down.

"Thank you, Jain. You are a wonderful person. It's always nice to know someone cares." LaDon smiled and they embraced once more.

LaDon headed toward the exit. He paused to take one last glance into his barren office. With a final sigh, he journeyed back the way he came. He reached his transport and directed it toward Nalkalin.

He arrived at the security gate and handed his Holopass to the guard. They recharged it with his valid credentials. He glanced at his Holopass after the guard handed it back to him. Anything techy always received a little admiration. LaDon was always a sucker for the finesse of technology. As he reached the lift, he stepped inside and pressed the glowing blue button labeled OD. He smiled because this time he knew what it meant. The lift doors opened. He could faintly hear familiar voices coming from the other side of the door. No doubt the voices of Jendall, Phelix, and Larissa. He opened both doors and saw their faces all staring at him. They each smiled and nodded as they recognized his presence.

"LaDon! So you came back huh?" Jendall said with a grin as LaDon caught Larissa in mid eye-roll.

"Of course. Expect nothing less." LaDon answered the senseless joke as best he could.

"We didn't know if you would be coming up here or working from your office," Phelix said as LaDon stepped down to the main floor.

"My office?" LaDon asked. "You mean my old office? The one they cleaned out from wall to wall?"

"Ah, not exactly. You see, yesterday you must have come straight from the Unification Chamber to the Observadome. Since you are the lead on this project, you have an office. There might be four Solasphere stations in the Observadome, but you also have your own. It's in your new office. You had a terminal in your old office, I presume. As a matter of fact, I took some Solaspheres to your new office this morning," Larissa explained.

"Are there any other surprises today?" LaDon asked while trying to avoid a cynical tone.

"With these two around...humph..." Larissa said as Jendall and Phelix ignored the comment.

"Well, I'd like to see this office, if you don't mind. Also, I will need to message the Assembly to find out what they did with my belongings," LaDon said with a bit of irritation.

"Right this way, sir," Larissa answered as they exited the room and entered the lift.

It perturbed LaDon that his personal items were packed up like he'd passed away. Some things you should be allowed to do yourself. *What's done is done, I suppose.*

The lift stopped at the floor which contained all management staff within Nalkalin.

"The management floor? Really?" LaDon gasped.

"Of course, boss. Nothing but the best." Larissa smiled as she led him in the direction of his new office.

"Um, we're going to have to work something out. You know. About you guys calling me boss and such. I prefer LaDon, but, um..." LaDon looked for the end of his sentence.

"No problem. Jendall and Phelix know when to be formal and informal. I was simply trying to make you feel uncomfortable," Larissa said with a smile as she glanced over her shoulder at LaDon.

LaDon couldn't help but notice Larissa's smile. It drew him in and gave him butterflies in his gut. Not only was it the most beautiful smile he had even seen, but he could also see friendship and admiration. He knew then that he was attracted to Larissa, but he had to keep focused on the task at hand.

They arrived at an office. The door was closed.

"Your Holopass should allow you access." Larissa gestured toward his right shoulder.

LaDon removed the Holopass from his shoulder and waved it in front of the plate next to the door. The light turned pale blue and the door clicked. He pushed the door forward. This triggered the office lights to come on automatically. He exhaled heavily.

With a rush of relief, he saw all of his things neatly stacked on his new desk. *Ah, there it is. Wait, wow, this desk!* LaDon felt spoiled when he noticed the furnishings of his new office.

He walked around to the side of the desk and rolled his chair out of the way. He took the two boxes from atop his desk and placed them on the floor. He noticed the picture of Barton Urthorn wrapped neatly on top of the boxes. He was glad they paid special attention to his most prized office fixture.

"Is this...? No...", LaDon whispered aloud in disbelief.

He looked up at Larissa like a child about to open a huge birthday gift. He smiled and motioned for her to come around behind his desk. With a look of curiosity, she walked slowly around the desk. LaDon looked around for the activation switch. It was right where he expected it to be. He pressed the button and stepped back so he wouldn't miss anything. The desk began to whir. The complete backside of the desk broke apart. Four viewing screens made their way upward from inside the desk. They rose slowly into view. Their final resting place was just eye level with the occupant. The control panel appeared as a holographic image on the open surface of the desk. It was just as his design had specified. This was his idea from long ago. He had sent it to the Assembly for assessment and never heard anything back.

As he slowly recalled his design, he remembered one important element. "If I turn around and... No!" He gasped.

They actually built the entire thing!

He looked toward his chair. It glowed a faint yellow from the base of the seat. He pulled the chair up to the desk and slowly sat down. The viewing screens flickered as his body weight settled in the chair. The display read *Please Connect Solasphere When Ready.* He looked to the right arm of the chair and ran his finger across its surface. A small panel appeared with the same controls displayed in the hologram. It was like he was sitting inside his own head. To have his idea come to life was a feeling LaDon had never experienced until now.

"We will never see you again, will we?" Larissa asked with a smile as she stepped closer to get a better view of the desk. "It's the first one of its kind, isn't it?"

"It's brilliant! I mean, I don't mean I'm brilliant, I mean..." LaDon stuttered.

"I know exactly what you mean, LaDon," Larissa said as she placed her hand on LaDon's shoulder for reassurance. "I heard some folks talking about the design when they put it together. They mentioned it was yours. I couldn't wait for you to see it. Gosh, all of this secret keeping can drive a girl mad."

"I thought they had forgotten or dismissed the idea. I thought some of the mechanical requests were impossible," LaDon thought aloud to himself as he was still taking in the sight of such a creation. "You want to know a secret?"

"Do tell," Larissa answered in a gossipy tone.

"When I designed this station in my head,

imagining all the things I wanted the desk to do, there was one special thing I wanted most of all," LaDon explained building a little bit of tension in his voice.

With a few commands on his chair's control panel, the blinds behind him started to lower and darkness crept into the room.

Next, LaDon spoke into the air, "Lights!"

The room went dark except for the faint glow of the desk and its mechanics. The light from the desktop displays and underneath the chair caressed the room with a faint glow of blue and yellow hues. LaDon rolled back in his chair to take in the full effect. Larissa's eyes lit up in wonderment. The ambiance of colors filled her eyes. It was like nothing either of them had ever seen. Even in LaDon's imagination, he could not do it justice. It was simply a thing of beauty. The light found its way from the bottom of the desk, which added a warm effect that LaDon had not pictured in his mind. This only enhanced the light show.

"I imagined a dimly lit environment would add to the submersion effect when interfacing with a Solasphere. When I use the view screens, I can control the recordings from my desk panel, and when I am fully immersed inside the Solasphere, I can simply lean back in the chair. Either way, I have full control of the experience," LaDon explained as if demoing a new product to a prospective client.

"It is spectacular, LaDon," Larissa said matching LaDon's excitement.

"Well, I suppose we need to get back to the

Observadome and get started," LaDon said as he disengaged the desk.

The room slowly regained its original lighting. He pressed the button to the right of his desk. The view screens lowered into their hiding place, leaving a nice, empty workspace for future use. They both exited the office and started toward the Observadome.

LaDon began to think about what he had just experienced. He also flashed back to when he had originally designed the work station. *Why did the Assembly not tell me about the desk?*

Just as this thought crossed his mind, he decided to send a message of thanks to the Assembly. As he and Larissa entered the Observadome, he excused himself, and told her he needed to send a message. His eyes drifted into his viewer.

"Ms. Sonne has shown me my new office. I am honored. Also, I wish to thank you for the wonderful gift! It's everything I imagined and more. And I thought the Observadome was amazing!"

signed,

LaDon Grafter

He sent the message and entered the Observadome. LaDon immediately put his mind to work. He began to formulate today's plan. He started checking off things in his head. *the Assembly should be in the meeting with the Lead Representatives by now. Jendall and Phelix should already be mapping the time structure just outside the initial creation of the new universe. I'll begin recording the first stages of development.*

He sat down at his new terminal and spun around to the team.

"All right. Jendall, Phelix, how far have you gotten on—" LaDon was interrupted by his viewer.

It was an urgent message from the Assembly instructing him to report immediately to the Unification Chamber. Looking at the faces of the team, he saw their eyes lost in their viewers. Obviously, they had received the same message. Each team member read the contents of the message. Almost in unison, they looked at each other and stood from their seats.

"Well, it looks like there are more surprises after all," Larissa said in a serious tone and looked in LaDon's direction.

"It appears so," LaDon agreed as he turned toward the door.

Phelix, Jendall, and Larissa all followed him to the lift.

"Right now they're having the meeting with the representatives, aren't they?" Jendall asked as the others nodded and agreed. "But that means?"

"Yep. We're going to be in there too, so it seems," Phelix responded.

The silence in the room was deafening. With a few sighs and shifting in seats, LaDon could tell everyone realized the gravity of the situation. No one had ever been asked to join a meeting with the heads of the entire planet.

As they exited the lift, they marched down the hall like soldiers reporting for duty. None of them knew what to expect as they approached the meeting chamber. LaDon noticed even Larissa hadn't been informed of this request. She seemed just as concerned as everyone else.

LaDon reached for the door, opened it, and stepped aside to allow his team first access to the room. He heard concerned, familiar voices coming from inside the room. As LaDon stepped inside, he saw all of the Lead Representatives arranged in their seats. the Assembly occupied their respective places, just like a Solasphere recording. This time there were three seats just to the left of the representatives. These chairs were filled with faces he did not recognize right away. The expressions on their faces looked as if they had each seen their own personal ghost. Blaine stood from his seat to address LaDon's group.

"Please, if you will take the seats that have been made available to you," Blaine said with his usual demeanor.

LaDon noticed chairs for each of them to sit. The team made their way to their seats. LaDon was the last to reach his seat. He noticed the representatives had paused their chatter momentarily to observe the new guests entering the room. LaDon noticed Joh Lin looking at him. LaDon smiled when he met Joh Lin's gaze. Joh Lin smiled politely and turned his gaze back toward the Assembly.

"All right, we need to bring the newcomers up to speed," Blaine said quickly.

The room was silent as Blaine took the floor.

"During our meeting with the representatives, we were interrupted by our friends from deep space. We all know they have many fields within their own area of study. Obviously, their main focus is the exploration of deep space. They also focus on the formation of our own universe. Finally, some focus on our own galaxy. The team sitting here today focuses specifically on our galaxy. They have come to us with news that we cannot ignore. In short, we had been led to believe that our sun had thousands of years left in its life. This allowed us time to explore deep space and look for an alternative planet to inhabit. Well, it seems we do not have as long as we'd originally speculated. New readings have shown that the decay of our sun will start to affect life on this planet two hundred years from now. As the sun starts to lose its mass, life as we know it will slowly deteriorate. Lack of energy, lack of warmth, and climate changes beyond measure. Deep space exploration is limited to the speed in which we can currently travel. We feel we do not have the time we need to find an inhabitable planet. We must find other alternatives. This is why you have been brought to this meeting," Blaine paused to let LaDon and his team take in the news. "We know the science is brand new. It's new to us all. But we have been speaking about this all morning and have concluded that this is the best course of action."

"Pardon my ignorance sir, but what do you mean?" LaDon asked with a puzzled expression.

"Compared to the time it takes for us to reach the closest solar system, your team can instantly place a Solasphere anywhere you'd like in this new universe," Blaine explained to LaDon.

This gave LaDon more questions than answers.

"Sir, if I understand you correctly, you are saying that we could find an inhabitable planet in the other universe faster than deep space explorers in ours. While this makes perfect sense, that would mean you intend to transport living tissue into this other universe in order to evacuate the planet," LaDon said aloud as he processed the information into a workable form.

Barton leaned forward in his seat, "Exactly, Mr. Grafter. We have already sent word to start working on such a plan. We have our best people on the job. Solaya is in danger, and we must protect our species."

"I understand completely sir. So what do you need from us?" LaDon sat up in his seat, readying himself for the responsibility.

Blaine answered in a slow, even cadence, "It's simple. We need to find an inhabitable planet. Specifically one inside your new universe."

"Sir, who's to say an inhabitable planet exists in this new universe? Furthermore, we have yet to understand its physics or its mathematical principles," Larissa inquired as she was trying to wrap this new information around her brain.

Alex sat up to address Larissa, "It's our only alternative at this point. We will continue to fund deep space exploration, putting all available resources toward finding a planet near to us. But we feel your project may hold the key. We are going to approach this from different angles. We will also need to go with some assumptions. We must assume the new universe behaves similar to ours. We will have to assume gravity behaves the same as it does in our universe. Also, all other mathematical equations compare with our own. Now LaDon, we know from your report yesterday that you were going to start mapping the beginning of the universe. This approach will need to change."

"I realize what needs to be done. I will prepare my report and explain our method of approach before the day's end." LaDon answered for his team.

With a firm pat on the back from Phelix, LaDon could see his team seemed relieved that he already had a plan.

"Very good, Mr. Grafter. We look forward to it. Representatives, you will each receive this report in the morning once we have reviewed it." Barton responded as he turned to face the deep space team. "As for your team, we want a report on the status of all deep space explorers currently in transit. Have them report their positions and hold for further orders. Good luck to you all. Both teams are dismissed."

Chapter 9

To Fail to Plan is to Plan to Fail

The entire team knew their task. They exited the meeting and quietly walked back to the Observadome. Although full of ideas, each of them was silent. The impact of the news had them a bit distracted. As they reached the Observadome, LaDon plopped down into his chair, leaned back, and looked toward the ceiling. *All of Solaya's history, destroyed in two hundred years? Two hundred years doesn't seem like enough time to chart an entire universe. Let alone using a brand-new technology that hasn't even had time to stretch its legs. Either way, failure is simply not an option. We must preserve our species at all costs. We can make this work.*

LaDon finished his inner monologue and sat up with confidence. Now he was face to face with the doubting expressions of his new team. What could he possibly say to the faces that were staring at him? They were obviously looking for hope and inspiration after hearing that their planet's fate rested in their hands?

"I know we received some very depressing news just now. I am having a hard time believing this myself. Nevertheless, we must remain focused. I suggest we continue our work as planned. We will start by learning how time passes within this universe in relation to the time variable. In the meantime, Larissa, I need you to be in communication with our friends from deep space," LaDon explained with a determined face.

This seemed to get everyone's mind off their impending doom for a short moment.

"What should I discuss with them?" Larissa asked somberly.

"We need them to tell us the age exact of our planet. All the way to the exact year if possible. I realize this is all theoretical at this point, but we need something to go by. Agreed?" LaDon looked to everyone for approval.

The three scientists nodded as they focused on LaDon for more information. It seemed the more he talked the more they could concentrate on the task at hand.

"Ah, I understand. I will get on that right away." Larissa didn't move from her seat.

"Jendall and Phelix, I want you to..." LaDon stopped for a moment to take a deep breath, slowing his voice to show his emotional concern. "Guys, I am just as shaken up by all of this. I am scared too. But in the end, there's nothing we can do about it except use what we have at our disposal. Look, this planet has overcome obstacles for centuries simply by building something better than what they already had. Necessity has been the mother of invention for this planet since old recordings of the Forgotten Wars. This is just another challenge. We may be a new team, but we are a team nonetheless."

"But we can't even find an inhabitable planet within our own universe. What's to make us think we will find one in this new one?" Phelix flung himself back in his chair with a hopeless sigh.

"Yes, Phelix, but we have an advantage that my friends in deep space exploration do not," Larissa explained, her voice filling with optimism. "We can go anywhere, at any time, within this new universe. We have to travel light years to get anywhere in our universe."

"That's true, Phelix. Think about it for a second." Jendall explained starting to get on board. "We can each track our own galaxies. We can watch the progression, skipping through time as it develops. We can calculate the patterns and simple mathematics to determine whether that specific galaxy might ever contain life. We could each go through numerous galaxies per day."

"Yes, I get where you're headed with that!" Phelix exclaimed as if he was making a brand-new discovery.

"See, you guys came up with that without me. I bet it just might work." LaDon said with confidence.

"Let's look at one galaxy together. This way we can come up with the best approach. First, we must understand how much time passes between whole numbers. We still haven't got that, yet. Let's set the time variable to one and place a Solasphere well outside the path of the initial explosion. We need to set it very, very far away as the explosion will be traveling at light speed. We can measure the rate of expansion that way." LaDon reached for the first Solasphere.

Larissa began her communication to the deep space team. The rest of the team followed the steps laid out by LaDon. They discovered some groundbreaking news within the first two hours of their research. LaDon stopped immediately once he made his initial calculations.

"Phelix, Jendall, please join me here a moment. I believe I see how fast time is moving compared to the actual time variable. I have been following this particular galaxy through space for a few hours and it seems for every one thousandth of the time variable, a year passes. I'm calculating this by the timer on the Solasphere display. It should be accurate enough information in my opinion." LaDon showed Phelix and Jendall his math. "Could you verify these numbers for me? I would feel better if someone checked my math."

Phelix and Jendall immediately began their own calculations based on LaDon's findings. It didn't take them long to agree that LaDon's reasoning made sense.

"This equation here. It seems to represent motion of some type?" Phelix's voice rose on the last word.

"Yes. I learned that I must place the Solasphere at the right moment to watch the galaxy zoom past at a tremendous rate. Remember, the galaxies are moving through space rapidly," LaDon explained. Both Phelix's and Jendall's eyes lit up.

"After breaking several Solaspheres, that sure

explains a lot. Scientist, I am. Astrophysicist, I am not." Jendall laughed a bit at his own humor.

The team worked tirelessly through the day and into the night. Solasphere after Solasphere, they continued methodically mapping the landscape of this new universe. All the while, noting which galaxies might have the possibility of life at some point. LaDon was getting ready to call it a day and let the team get some rest when Larissa spoke up from her station.

"Guys, I realize it's a little late. But I just received word from our friends in deep space concerning the birth of our galaxy. They also have some good estimates on what to look for after the initial explosion. They say to start about nine billion years after the explosion. They also have notes showing how far a planetary body should be from the star it orbits. This gives a better chance for the planet to be inhabitable by a species similar to our own," Larissa informed the team as she forwarded her information to their viewers.

"Wow, that's a lot of information! Good information, no doubt. Although, I can see us all getting a bit tired. This is perfect place to stop. We can start fresh in the morning, agreed?" LaDon asked. The team agreed without hesitation.

Everyone gathered their things. LaDon began thinking about the new information Larissa had just delivered to his viewer. He had begun to drift into his view screen when he noticed a presence next to him.

"Hey, LaDon?" Larissa's voice asked softly.

"Oh, you startled me a bit. Yes, Larissa, what's

on your mind?" LaDon answered with his best professional voice.

"We were thinking about having a quick meal before heading home. I know we've been working hard and probably need to rest, but I don't see us getting much sleep after the news we've received today. Please, come with us. Remember, you promised," Larissa pleaded in a tired voice. "Don't leave me alone with them."

"Of course. I would love to," LaDon answered with a tired smile.

LaDon knew he needed to spend time with his team outside of work. Besides, he hadn't even seen his other friends since starting this assignment. At this point, he felt he might as well make new ones.

It didn't hurt that Larissa Sonne invited him. When she startled him, he noticed just how close she was standing. It made his heart begin to pound and his thoughts start to scramble. He had never been so awestruck over someone in his life. There was just something about Larissa that he found irresistible. His brain stumbled each time she addressed him. He knew he must keep the relationship professional, but his mind made every excuse to circumvent social restrictions.

LaDon and Larissa began walking toward the exit when he noticed Jendall and Phelix catching up to them.

"So, she talked you into going, huh?" Jendall asked with his usual happy tone.

"She didn't have to convince me. I'm more than happy to join you," LaDon explained as he shot a small glance in Larissa's direction. "I'll be returning for a short period of time after we eat. I have a few ideas I want to try."

LaDon and Larissa exchanged glances. Jendall pushed between them, reaching for the door. They filed out of the Observadome and headed toward their respective transports.

"So what are you wanting to work on, LaDon?" Larissa asked.

"I want to estimate how long it will take to pinpoint a galaxy. This will help me plan our schedule for tomorrow."

Larissa grimaced in frustration. "You say we need rest, yet you continue working. How can you explain your new plan if you haven't slept?" Larissa asked sternly.

"I know, I know. It's just the way I've always been, especially now, knowing what we know," said LaDon as his words trailed off. "Plus, I get to play with my new desk."

"Well, at least you're coming to dinner. Besides, you could probably use a little distraction." Larissa's eyes met his, and they both smiled.

"I agree," LaDon replied as he broke their gaze.

He could still see her smiling from the corner of his eye.

She's probably just being nice. Get a hold of yourself. Can't someone just be nice and not have ulterior motives?

He started to realize he was on the verge of a complete freak out moment. He attempted to be more aware of his demeanor and decided he was containing his emotions quite well. After a few inward slaps in the face to slow his adrenaline, his nerves calmed a bit. *You're just going to dinner with your coworkers. That was not an advance. Don't lose it.* As he began to calm down, he wondered what had come over him. Maybe it was the way she said the word *distraction* that caught his attention. Either way, the only conclusion was he was developing feelings for Larissa. After a little time, he could slow his heart rate and think clearly. *Hopefully we can find things to talk about at dinner. I hope I don't sound like a fool.* Suddenly, he realized how silent the conversation had become while he was stuck in his thoughts. He thought quickly for something to say.

"So where are we going? Is it the same place you all went yesterday?" LaDon asked as he listened for his own voice to make sure it was steady.

"Yes. Neither Jendall nor Phelix like change. Plus, it's close to both of their homes. I just go with the flow. They're fun to hang out with. Phelix is always level headed. Jendall needs to take something to calm him down at times. You know, I think he forgets where he is sometimes. He can say the craziest things when the mood catches him right. Oh no, I'm rambling." Larissa cut her words short and paused for LaDon's feedback.

LaDon agreed, "Well, so far I tend to agree. They are both diligent at their work and the top minds in their respective fields. Not to mention quick of wit. They are a joy to be around."

LaDon and Larissa stopped when they heard the voices of Jendall and Phelix from a few paces ahead of them.

"Hey, we are going to The Stamosh again, right?" Jendall shouted from across the void.

"Yeah, that's the plan." Larissa waved at them as they exited Nalkalin en route to their transports.

"The Stamosh it is then," Phelix called.

Larissa directed her attention back to the conversation.

"So, what about me?" Larissa asked shyly.

"What? Um, how do you mean?" LaDon asked obviously avoiding the question.

"I said, what about me? You've told me what you think about those two crazy Solayans. Now, what do you think about me so far?" Larissa asked again, slowing her words and making deliberate eye contact with him.

"Well, um...you are...", LaDon noticed his tongue stumbling around for the right words.

"Ha! Put you on the spot, didn't I?" Larissa laughed as she nudged up against him.

"Well, no, not exactly. I mean yes. I mean, to answer your question, I would say you help me keep my sanity in that room. Like when they start to ramble about linear time and equations. You know, I believe on some level they think I'm a world-class scientist with equations floating around in my head," LaDon joked.

"Like I said, Jendall's mind is so unpredictable at times, and Phelix seems to take those little trips with him. Somehow, strangely enough, they make a good team," Larissa responded. "But you didn't completely answer my question."

LaDon cringed inside as he realized he had not escaped the question. She really was putting him on the spot. If he answered the question truthfully, he was not sure what her reaction might be. As they approached the parking area, LaDon came to a stop next to his own transport.

"I'm parked just a few spaces that direction," Larissa said as she stopped at LaDon's transport with him. "I'm not moving until you answer my question. You told me your assessment of them. Now I want my assessment."

With a sudden rush of newfound confidence, or maybe it was sheer nerves blocking his cognitive reasoning skills, his true feelings suddenly spilled from his mouth.

"I think you're brilliant. You are very aware of what goes on around you and you enjoy what you do." LaDon said as his heart began to race a little faster.

"Is that all?" Larissa closed the space between them with a small step.

"Um, well, that's all I can say being your boss." LaDon smiled, looking for every reason to break the rules.

"So, there's more? Oh, now you've got me curious," Larissa said with the same irresistible smile LaDon had been battling the last couple of days.

In less than a second, LaDon let the last few days flash through his mind. He had never experienced such change in a short amount of time. A bit more wasn't going to hurt. He had been able to resist her charms up to this point. Their communication had been nothing more than shameless flirting from across the room. Now, with Larissa outwardly showing the same feelings he had been fighting, he decided he could use that distraction she mentioned earlier.

"Well, I think you're stubborn, for one." LaDon closed the gap between then even more. "But that's obvious, otherwise we wouldn't still be standing here."

"Go on." Larissa smiled and blushed as she folded her arms.

Finding that same rush of confidence he felt moments ago, LaDon gave up his inhibitions.

"At risk of being unprofessional, I think you are an amazing woman with a tantalizing smile," LaDon said as he closed the final bit of space between them.

Larissa's arms unfolded, allowing LaDon to step closer. He leaned forward. Their lips met. LaDon wrapped his arms gently around her waist. As he pulled her closer, she delicately placed her hands on the sides of his arms. It was tough, but he made himself stop. He did not want to overdo it. As he moved back, he noticed her eyes were closed. This gave him immediate reassurance that this was exactly what she hoped would happen. His mind raced as he looked for the right words to say. With his mind tired of thinking everything through, he kissed her again. This time she was more than receptive. He could hear her breathing intensify. This kiss was more passionate than the one before. LaDon drew back again. He wanted to see her beautiful face. Again, her eyes were closed, as if she was yearning for more. He could not believe a girl as beautiful as Larissa Sonne had allowed him to kiss her. This was something he definitely didn't plan for. LaDon was always in control of his day to day routine. He never saw this coming.

"I am going to take this as a good assessment," Larissa said softly as she slowly opened her eyes.

"I would say that's accurate." LaDon smiled as he caught his breath.

LaDon noticed they were still holding on to one other. He wished time would slow down and this moment would last just a little longer.

"I might stand here like this all night. But I suppose we should catch up with them," LaDon said reluctantly as they relaxed their grip on each other.

"Yeah, I would say so. I'll never hear the end of it if they suspect anything." Larissa sighed as they continued to untangle themselves.

Larissa took a few steps back, slowly dropping LaDon's hand. She turned and headed toward her transport.

With a smile, she looked back and said, "If you're not sure how to get to the restaurant, just follow me."

"I'm right behind you." LaDon lowered himself into his transport and closed the door.

All at once, LaDon's insides shuddered with

giddiness. He felt like running in one direction as hard as he could. He let out a deep sigh and attempted to push out the jitters. It worked momentarily, but nothing was going to remove this angst until he kissed her again. *Who knows, maybe I was dreaming, right? Hah! No way.*

Chapter 10

Here's to New Friends and Then Some

The hum of the engine cleared LaDon's mind. He was definitely not dreaming. He could still smell the faint aroma of her perfume and taste the sweetness of her lips.

Coming back to the here and now, LaDon remembered he needed to follow Larissa since he didn't know where the restaurant was. He watched behind him for Larissa's transport. She drove past and waited for him to follow.

He stayed close and reflected on what happened. He immediately began having doubts whether or not he should have a relationship with a co-worker. Not to mention someone who reported to him. This went against everything he had been taught. The other side of his brain, the one connected to his heart, disagreed immensely. LaDon had never met a woman with such audacity and beauty. He had always believed these two traits to be an impossible mix until he met her. He figured they were both adults and should be able to handle the situation.

They arrived at the restaurant at the same time. Jendall and Phelix waited patiently on a bench outside the entrance to the restaurant. LaDon found a parking space, disengaged the engines, and took deep breath. He attempted to push away the anxiety brought on by his encounter with Larissa. LaDon exited his transport.

Larissa had already made her way to the bench where Phelix and Jendall were sitting. He could already imagine the onslaught of questions from Jendall being aimed at Larissa. He couldn't quite make out what they were saying, but she was laughing. Jendall was obviously up to no good. It brought a smile to his face to see the chemistry between his new team. As he approached the group, it was Phelix's voice that LaDon picked out first.

"Great, you're both finally here. I'm starving!" Phelix said as they all turned to enter the establishment.

Surprisingly, neither Jendall nor Phelix mentioned anything about Larissa and LaDon's slight tardiness. At least nothing LaDon had heard. On one hand, this surprised LaDon. He expected Jendall to jump at the chance for a joke. On the other hand, it pleased LaDon to know his new team could be professional even when not on the job. They were led to their seats without waiting. They were seated quickly since the restaurant was half empty.

The waiter looked to Jendall and Phelix. "Same as before?"

"Yes," they both mumbled in unison.

"And you, ma'am?" The waiter turned toward Larissa.

"Oh, I'll have a Pumosh please," Larissa

"And you, sir? I don't believe I've ever seen your face in here with these characters." The waiter smiled and waved a finger in Jendall and Phelix's direction.

"Ah, yes. We just started working together. Let's see. Being that I've never been here before, I'll try whatever he's having." LaDon gestured in Jendall's direction.

"You sure? It's a little strong," Jendall replied with a side-mouth smirk and a raised brow.

"I can take it. We've had so much thrown at us all at once, I feel we all need a little slap in the face. Don't you agree?" LaDon smiled as he scanned the faces at the table.

The meal arrived and conversation ensued. Attempting to defog his mind of thoughts about Larissa, LaDon took the time to focus on each member of his team. He wanted to learn each of their stories.

As the night unfolded, he learned that Phelix was married with no children. His wife ran a local bakery close to their house. LaDon knew the bakery well. They used to deliver pastries to his old work place. *Such a strange connection,* LaDon thought to himself. Larissa lived alone and had a very close relationship with her family. She spoke often about her mother and father and how they got her interested in deep space. Finally, he learned that Jendall was a family man who loved his children very much. His wife was a teacher at The School of Science and Reasoning. She headed up the exams for those that had reached The Age of Understanding. LaDon reflected on the day he took his exam. It wasn't very difficult. Most of the test was common sense and knowledge about their planet's history. Being the grandson of the famous Pomph Grafter, he definitely had an advantage on that part of the exam. As his mind came back to the present, he listened as Jendall explained the purpose behind the test.

"You see, the exam is not designed to test the *aptitude* of the applicant. The test is used to measure the essence of the Solayan. It is designed to determine whether or not your common sense is worthy of moving forward in society. If you can't pass, you can't gain viewer implants," Jendall explained, showing his firm belief in the guidance of the Assembly and their teachings.

"It thins out the misfits, you mean," Phelix answered in a blunt tone. "I'm glad it's there, personally. There are some really fouled up people that need to be culled until their brain can fully develop."

"Phelix, was that an attempt at humor?" LaDon teased

"I believe it was," Larissa exclaimed, confirming LaDon's theory.

Phelix looked down at his plate with a sheepish grin as his complexion flushed red. "There's probably just something in my drink."

They all laughed together at the rarity of Phelix's attempt at humor. The laughter subsided and idle chat resumed.

The night was filled with talk of home and

family. They didn't talk much of the pressing news they received earlier in the day. The dinner was a much-needed escape. As the meal ended, LaDon decided the time together was an excellent idea. He was glad he'd decided to come along. LaDon noticed that Larissa had not made much eye contact since they arrived. It gave him a sense of relief to know she was as mature as she looked. The evening came to a close as they each paid their tab and weaved between the tables toward the exit.

"Have a pleasant evening and thank you for inviting me to come along. I have enjoyed the company," LaDon said to them as they move toward their transports. "I'm going to go drop a few more Solaspheres at some particular locations. Just a hunch."

"Don't overdo it, LaDon. Just remember to get some rest," Phelix advised from a distance as they each waved farewell.

LaDon walked toward his transport, and Jendall and Phelix headed in the opposite direction. Their transports must have been parked on the other side of the establishment. He passed Larissa's transport on the way to his. This meant Larissa must be right behind him. His heart raced as he reached his transport. He was hoping beyond all hope that he would turn around and see her standing there. Another part of him was in absolute terror, hoping she was already driving away. He listened for footfalls and heard the faint sound just behind him.

Suddenly, he heard the whir of a transport. He looked to his left and saw Jendall and Phelix, one after the other, waving farewell. LaDon returned the farewell gesture. From the corner of his eye, he saw her. She had just passed her transport and was headed in his direction. He couldn't help but smile a bit knowing she was coming to see him. He turned to face her. She smiled shyly as she approached, seeming a bit off kilter. LaDon wondered, *Oh no, does she regret what happened?*

"Hi," Larissa said, almost as shy as she looked.

"Hey. Thank you for inviting me tonight. I had a wonderful time," LaDon said, breaking the ice perfectly. "Tonight was a real eye opener for me. You know, I feel with our planet facing inevitable destruction, this team is the best one for the job. I don't know what it is, but something inside me feels like this team was organized specifically for this task. Almost like fate. Do you believe in fate, Larissa?"

"Fate isn't something you can measure. When searching for a solution, you can always find one...if you don't die first. Anything is solvable if you have enough time. Our history is living proof. During the Forgotten Wars, one invention led to another invention which led to another. To call this fate implies an outside force or entity imposed its will on the outcome. For me, history is measurable; therefore, its repetitiveness is what I believe. What we learn from history makes us better today. As for fate bringing our team together, I call it coincidence," Larissa answered with obvious conviction.

LaDon was struck speechless. He could not believe the words he had just heard. It was like

Larissa had been put on this planet just for him. Never in his life had he been more attracted to a woman than he was at that moment. It was like she had snuck into his mind and repeated his thoughts back to him. What started as physical attraction just became much more in LaDon's mind. If only he could let her know how deep those words ran without scaring her off, he would have told her right then. He thought of the most obvious response. A response that would not seem like he just fell head over heels in love.

LaDon reached out for her hand, which she warmly accepted and said, "I couldn't have said it better myself."

"Oh? I thought you were about to tell me you believed in fate," Larissa replied.

"No. Not one bit. It's almost like you were reading my thoughts. It was quite wonderful to hear actually," LaDon answered as a small sigh slipped out.

"Well, I guess I need to be heading home. It still bothers me that you are going back to Nalkalin instead of getting some rest. But no matter what I say, you're going anyway, right?" Larissa said with a knowing smirk.

"How about this? I promise I will not stay long. I just want to try one thing."

"Just keep in mind these next few days, we are going to need you clear headed. But please don't think I am telling you what to do. I'm just thinking about you," Larissa said almost lovingly.

"I understand completely. I promise not to be

long," LaDon answered.

"Well, don't let me keep you from your 'one thing.' Goodnight LaDon." Larissa smiled as their hands finally dropped.

She turned to leave, but LaDon didn't want her to leave without somehow addressing what occurred just before they left the parking deck. *She has got to be thinking the same thing, right?*

She started to turn toward her transport when LaDon softly whispered her name, "Larissa?"

She turned around instantly and their eyes met, "Yes?"

LaDon took a few brisk steps toward her. He wrapped his arms around her gently and moved his face closer to hers. LaDon was flooded with the aroma of her perfume and the warmth of her body as their lips met. Larissa leaned into him. He was filled with excitement. Now he was certain she did not regret their first encounter. She was obviously enjoying the moment. As the kiss came to an end, their hands found each other and clasped softly.

"You'd better leave before I take you home with me," LaDon said as they backed away from each another.

"I was worried that you regretted what happened in the parking deck," Larissa explained with her eyes focused on LaDon's.

LaDon laughed. "I was thinking that about you. Obviously, we both were wrong."

LaDon took a few steps back as she turned back toward her transport. She reached her door and looked in his direction.

With a small wave, she flashed her amazing smile once again.

"Goodnight," she said sweetly.

"Goodnight," LaDon said as his emotions ran wild.

He entered his transport, watching her the entire time as she drove off into the distance. With her completely out of sight, he tried desperately to divert his attention to the task at hand. No matter how hard he tried, he couldn't quite remove the thoughts of the evening from his mind's eye.

***

LaDon arrived at the gates of Nalkalin. At such a late hour, once again, LaDon couldn't help but notice the beauty of Nalkalin. This was the first time he had gotten to experience the building at night. Most Solasphere recordings occurred when everyone was awake. The structure looked like a child's room lit up with a night light for comfort. He pulled up to the guard station to have his Holopass credentials reassigned.

"Late night, sir?" asked the security guard.

"Yep. Just one more thing, then I'm headed home," LaDon answered as the security guard passed LaDon his Holopass.

"Don't work too late, Mr. Grafter," the security guard said as he engaged the gate to allow passage.

LaDon pulled into the parking area at the same location as his first encounter with Larissa. He played a small movie in his mind of what had happened. He smiled a bit as he disengaged the engines. Silence filled the transport like fog surrounding a mountain top. He tilted his head back on the head rest and closed his eyes. He took a moment to remember the softness of her voice and the sparks of passion they experienced during their brief moments. Realizing he was drifting away, he shook it off with a deep growl and climbed out of his transport.

He had a bit of giddiness since this would be the first time in his new office. He'd get to be alone with the state of the art viewing station he had designed. All of the lights on his floor were off. He made his way through the dimly lit corridors, finally reaching his office.

LaDon slipped to the back of the desk and reached for the activation button. The desk flared to life and the glow from the desk slowly lit up the room. Wonder and amazement filled his mind. Now he could fully enjoy the ambiance and marvel at the intricacies of his new toy. LaDon took a brief moment to activate the network of new information being recorded. Working in the Observadome had kept him out of the loop of current events. In his mind, this was simply unacceptable. He put in his credentials to see the latest information stored over the last few days. He saw nothing of interest. A few police disputes, one presentation of an award for outstanding achievement at a sporting event, and a few recordings that may be deemed important on a later date. LaDon felt good about the fact that he hadn't missed much in the world of Solaya while he had been tromping around in the other universe.

With the present day news safely stored away, LaDon reached down to retrieve one of the Solaspheres that was brought up to his office. *Nine billion years, huh? Well, let's just see,* LaDon said to himself as he set the coordinates of the first Solasphere. He connected the device and placed it inside the casing. The same flickering blue light appeared, just like in the Observadome, and poof. The device disappeared. Another flash and it returned.

He continued to send a few Solaspheres, viewing each one upon its return. He decided to pick a distant star and work his way toward it. Solasphere after Solasphere, he ran the calculations through his new terminal and finally got within range. *This wasn't difficult as I first imagined. I reached this star in a matter of minutes with these calculations. If I adjust for the rate this particular galaxy is traveling, I can follow its birth and record it from a distance. Heck, I could track it all the way back to the first initial explosion of the universe.* That was exactly what LaDon was hoping to accomplish that night. Find a galaxy, study its path as it travels through the new universe, and record the galaxy's life as it evolved. LaDon took a few more minutes watching Solasphere recordings, figuring out how difficult it would be to watch for patterns matching their own solar system. He paid close attention to which planets orbiting the sun might possibly sustain life. About thirty minutes had passed before he reached the conclusion that this particular system did not contain life. *Such a marvelous invention to be used for such desperate measures.* LaDon disconnected the last Solasphere and pressed the power switch. The terminal morphed back into the desk it was when he first walked in the door. He walked back to his transport, engaged the engines, and headed for home.

During the drive, images of the day passed before his eyes. Images of the new universe, Phelix, Jendall, and the Assembly all rushed through his mind simultaneously. His mind of course drifted toward Larissa. Such piercing eyes, stunning straight hair, and lips that could draw the tension from his body like a siphon. *How on Solaya can she be so beautiful?*

Chapter 11

Where Are You Little Star?

Images of the night before flashed through his mind like an old movie reel. Still, LaDon tried to enjoy the morning ride to Nalkalin. After passing through the normal security check, he proceeded to the parking area. Once again, Nalkalin was full of life as compared to the night before. Random Solayans scurried to and from their destinations. The parking area was now full of transports that weren't there just hours before.

LaDon found an open space. He noticed a few familiar transports. Both Jendall and Phelix were parked in their normal spots, and Larissa's sat just a few spaces down. *Gosh, they always beat me here,* LaDon thought with a chuckle.

He headed directly to the Observadome to share with his team what he learned the night before. Just as he reached the top of the lift, he received a message on his display. It was a message from Alex Cuberly, Time Keeper of the Assembly. One of Alex's many roles was to ensure all information relayed from the Assembly was properly delivered to the recipient. The message read:

LaDon,

I have some information for you and your team. It was sent directly to me from our friends on the deep space team. I feel this information will be vital to the next phase in your quest. Rather than send it over the viewer, I will be making a trip to the Observadome to deliver the news myself. I will send the information over the viewer once I have delivered the news personally. To be totally honest, I am fond of the new Observadome. I want to see it now that all construction is complete. I think the structure is an absolute delight. See you soon.

signed,

Alex Cuberly

Time Keeper of the Assembly

LaDon walked through the door to see the team working at their stations. Jendall was sipping his morning concoction. Phelix was propped with one hand on the back of Jendall's chair. LaDon could only see the top of Larissa's head. It seemed she was buried inside her displays.

The sound the door opening gave away his presence. They each broke their concentration momentarily to look up from their work. Jendall smiled with his normal joviality while Phelix calmly nodded in LaDon's direction. They both continued to their current tasks. LaDon looked toward Larissa, who briefly looked over the top of her monitor. She smiled pleasantly and eased back into the grip of her workstation. LaDon approached his terminal, arranged his things, and prepared to address the group. He quietly cleared his throat where no one could hear him and turned to face them.

"I hate to interrupt as you all seem deep in your thoughts, but I have some news to relay," LaDon announced, hating that he had to break their focus.

LaDon watched as his new friends politely looked up from their work and awaited the news from LaDon.

"I just received a message from Alex Cuberly. He will be paying us a visit soon. He wishes to relay some information from our friends on the deep space team. He says the information will be vital to our search. As an aside, Larissa, being from deep space research, would you happen to have any insight or foreknowledge of this information?" LaDon peered at Larissa.

"None whatsoever. I haven't talked to any of my friends from that area since we started this project."

"It must really be something good! I mean, for him to deliver this information personally. This is a real treat," Jendall said eagerly.

"He mentioned that he wants to see the facility now that the construction is complete. He probably thinks the same as I do. This building is definitely miraculous." LaDon's eyes followed the curve of the dome as he handed over the last bit of his information. "Until he arrives, I suggest we continue on with our plan. I do have one thing to share. I learned this last night after dinner. Again, I hate to interrupt your progress, but I feel this is important. Please gather around my terminal. I am going to take over the top quadrant of dome displays momentarily to show you the full scope of my idea."

The team disengaged their Solaspheres and locked their terminals. They seemed interested in what LaDon had to show them as he began sharing his findings with them.

"There was one thing I wanted to learn last night. I wanted to see how difficult it would be to pick a star and track its progress as it traveled through space. I approached it with a Solasphere and used the criteria shared by our deep space friends. From there, I was able to determine if or when this particular star system could be capable of sustaining life. The entire process took about thirty minutes for one star. I figure once we are good at this approach, between the four of us, we can cut this time dramatically. Look at these calculations for a moment." LaDon leaned back so everyone could see his console.

"Yes, very impressive. Jendall and I were just discussing the best way to track the actual rate in which the particular galaxy is traveling. These numbers might actually do the trick." Phelix said as he pored over LaDon's numbers. "Are you sure you're just a historian?"

"Hah! If there is anything I'm one hundred percent sure about, it's the fact that I'm only a historian." LaDon laughed as he accepted the compliment.

"I have a suggestion. What if we factored in the color spectrum of the star? This way we know ahead of time if the star is moving toward or away from our location. This would cut down the time spent getting closer to the star," Jendall said, which shocked LaDon as he had yet to hear the scientific side of the resident joker.

Larissa chimed in. "That makes perfect sense, Jendall! Actually that will cut a substantial amount of time. My only concern is simply the sheer number of stars. Around the nine billion year mark, there are just so many!"

At that moment, the team's conversation was interrupted by the sounds of a visitor entering the dome. They all looked toward the entrance to witness the arrival of Alex Cuberly. Alex smiled warmly and trotted down the few steps to arrive on the ground floor. He approached the team in order to share his information.

"Good morning, my fellow Solayans. I hope you have had a pleasant morning." Alex tilted his head back at some of the view screens. "This place is astonishing. You will have to show me all of its glory whenever there is a chance. Not now, of course. We have too much to do."

"I promise one of us will give you the complete tour, sir." LaDon answered, sharing in the admiration of the Observadome.

"Very well then. Down to business. the Assembly members figure your team is hard at work trying to decide the best way to go about searching for a suitable planet. I believe this bit of data will vastly improve your chances. This approach uses all the information we know about our own planet when viewed from a distance. Color, distance from our sun, age of our galaxy, and a few other variables. We figure if we use our own celestial surroundings as a guide, we can narrow the search field even greater. Yes, there are billions of stars out there, but this information should give us a fighting chance," Alex said with conviction as he awaited the team's reaction.

"That sounds great, sir. Let's see it," Phelix responded with equal conviction.

"I am transmitting it now."

The room fell silent as Alex sifted through messages in his viewer. He forwarded the news to the team. Each team member carefully read through the information.

Alex continued, "As you can see, they have approximated these figures based off our own planet. If we were to travel outside our own galaxy and look back, we would see certain things. I would like to point out one image specifically. It is the image labeled 'speck'. This image is an image of our planet as seen six billion miles away. It is simply a speck, but with the correct light refraction from our sun, we are able to see a dim, green tint. We suggest looking for this while utilizing other variables found by the deep space team. The color spectrum will lend itself to greens and blues if the planet were to contain life as we know it. I realize this one variable within itself seems farfetched, but if we use these ideas in tandem, it just might expedite the process you are using now. Does this make sense to everyone?"

The team was still busy digesting the information. This was their next phase of work already completed. Using this new knowledge along with LaDon's method, they could cover almost five times as many stars.

"If we cannot find an inhabitable planet before our time is up, then we were never meant to find one in the first place," Phelix said with astonishment in his tone. "I thought all of this would take days to determine."

"Well, while your team was getting used to the new Observadome, studying the layout of the new universe, and mapping the trajectory of the matter out into space, our deep space folks were working on this theory. They simply needed the information they had already ascertained about our own planet in order to provide this insight. One team working one side of the equation while the other side working the other. Our history once again repeating itself, eh LaDon?" Alex looked toward LaDon for a response.

"Quite true sir. A blatant display if I do say so myself." LaDon answered with a slight questioning tone. "It just seems so coincidental, sir. Almost to the point of disbelief."

With great passion, Alex began, "Then allow me to shed some light on your disbelief. You see, LaDon, the Assembly has a unique point of view when it comes to planetary evolution. We have the knowledge of the entire world at our fingertips. We meet with groups on a regular basis, learning what they know. Of course we cannot be experts on all of it, but we can see which heads might be the best to rub together. This has been the role of the Assembly for quite some time now. With this particular situation, finding this new technology that Jendall and Phelix developed is a grand find. In my opinion, it couldn't have come at a better time. With this knowledge readily available, we were able to face one of the greatest perils this planet has ever encountered. Our timeline, our history, is living proof that we can overcome grand obstacles. Even obstacles we create ourselves, such as the Forgotten Wars. We were such individuals back then. Now, we are united. If we fail, at least we go together, but at least we had a fighting chance. In our case, a fighting chance against extinction."

"I understand sir. I did not mean to sound doubtful," LaDon answered, subdued after such a riveting speech.

"No, you didn't sound doubtful at all. Think back, LaDon. Think back to all the Solaspheres you've watched and archived throughout your career. Countless meetings and endless questioning from representatives. Our perspective is one of a kind. We sit where we can see it all. Every facet of life. We're not just looking down a tunnel. We're looking out across the abyss." Alex said with conviction showing his enthusiasm. "Your questions are valid. I just want you to take a step back for a moment and see the big picture."

LaDon sat for a moment, placing his hands over his face while his team looked on. He began to reflect on the numerous Solasphere recordings throughout his life. He had always noticed how different Assembly meetings throughout history had somehow tied to some specific event later in life. He had never put two and two together until now. The same history explained by his grandfather in such detail had been the result of the Assembly's unique perspective the entire time. Alex was right. The bigger picture had been right there in front of him the whole time. Suddenly, LaDon looked up at Alex as if he'd seen a ghost. It's almost as if LaDon stepped away from his body, outside the Observadome, and looked down on his planet from above.

"What do we do when we find a planet?" LaDon asked with a shaky voice.

Alex leaned down, now eye level with LaDon. He seemed to beam with pride, proud of LaDon's epiphany.

With his enthusiasm about to boil over, Alex whispered, "Now you're there with us. And don't worry. the Assembly is already traveling that journey, my friend. Just find us a planet."

With a newfound confidence, knowing the Assembly was in their corner, LaDon sprang to his feet and said, "Then we have work to do."

The team agreed as they gathered behind LaDon. They were excited to apply this new information to their current process and start sifting through the numerous stars that awaited them.

"Then I will leave you to it." Alex slowly turned and exited the dome.

Before Alex could even get out of sight, LaDon started outlining his plan. "All right. Let's apply this new information and get a game plan before we begin. It's about an hour before we break for a bite to eat, so let's get a good rhythm so we will have something fresh to come back to once we have eaten. Does this sound acceptable to everyone?"

The chorus rang loudly with affirmative responses. LaDon and the team wove the new information into their current plan. Jendall and Phelix completed their analysis of LaDon's equation to chase down roaming galaxies. Making some minor adjustments in the math, they were able to perfect the process to a science. Next they applied the new information from their deep space consorts and began the search. After an hour, they had searched numerous galaxies. With no signs of life up to that point, they each broke from their terminals and wandered off in search of sustenance.

LaDon scanned his brain for the closest available food source, as he was anxious to search through more galaxies. He sat in his chair thinking of what he would like to eat. The room around him felt empty. This feeling dissipated when the air had changed behind him. He felt a gentle hand on his shoulder along with the fragrance of a familiar scent.

"You better eat something, sir. You won't be able to concentrate if you don't," said Larissa's voice behind him.

This whisked him away into his own world. Her voice was just as soft and pleasant as the night before. LaDon stood from his chair and turned to face her. He quickly scanned the room to see if Jendall and Phelix were still present. When he saw they were alone, he looked at her and smiled.

"I am. I'm going now. I've actually never been to the cafeteria yet. How's the food?" LaDon asked, keeping his emotions in check—kind of.

"It's worth trying. Shall we?"

"Lead the way."

He followed alongside Larissa as they traveled toward the cafeteria. Of course LaDon already knew the way. He had passed it on multiple occasions when floating through the halls in his Solaspheres. He even remembered the scents coming from the cafeteria when he would pass by the doors. He'd swear he remembered smelling his favorite dish a long time before. Larissa probably knew that LaDon would know the way. She was quick on the uptake, but the subject matter seemed too silly to mention.

After bouts of small talk and LaDon battling his desire to stand closer to her, they reached the cafeteria. During the meal, the conversation turned to talk of the galaxies they had just recorded, all the while making certain faces only readable by the two of them. It made his insides turn over on themselves. He wanted to discuss the night before, but he knew it was for the best. They were at work now. There would be plenty of time for that. He could tell she was probably feeling the same way by her eye contact, facial expressions, and constant blushing.

Lunch felt like it was over before it began. They gathered their things and headed back to the Observadome. Jendall and Phelix arrived almost at the same time. They all went back to work just as eager to begin as LaDon.

Several days went by with nothing significant to report to the Assembly in their daily reports. A few barren chunks of rock floating around large stars similar to their own sun. Beautiful light shows consisting of colliding galaxies and exploding stars. Radiant pulsars and even black holes topped the list of interesting topics of discussion. Unfortunately, there were no planetary bodies even close to meeting the requirements of life.

***

About one week into their search, LaDon settled into his chair. He knew they were all in for another day of star searching. LaDon looked to see the area of the universe his colleagues were exploring. He decided to start his search in a completely different direction, just to shake things up a bit. LaDon connected his Solasphere and began the arduous process. Galaxy after galaxy, LaDon was still mesmerized by the brilliant formations. He watched the birth of each cosmic landscape unfold before his eyes. It was tough at times to be subjective, looking for scientific data while such beautiful arrays of colors passed before his eyes. The brilliance of sunlight at the center of each galaxy seemed just as radiant as the next. He even heard gasps of astonishment from his team as they experienced their own wonders.

"Wow, you guys should see this," Jendall exclaimed in an obviously rhetorical tone.

"Unbelievable," Phelix commented at another random moment.

Each team member continued searching, stopping from time to time to rub their eyes and take sips of their beverages.

LaDon decided to take a small breather. He looked over toward Larissa. He saw her leaning back slightly in her chair, probably watching another beautiful solar system forming or dying before her very eyes. He saw her expression as she reentered reality. She sighed a bit as she disconnected the Solasphere. She looked over at LaDon, as if she felt him staring at her, and immediately flashed her stunning smile. He immediately responded with the same gesture.

The past week had allowed LaDon to become more and more comfortable with his relationship with Larissa. They both had worked hard to mask their feelings in the work place while stealing a glance here and there. He even got to spend some time with her over the weekend. It was a much-needed escape from their work. He had found her to be a true woman of beauty both inside and out.

"Don't get too frustrated too soon. There's billions of stars out there," LaDon said across the room.

"He's right, Larissa. We'll find one," Phelix added with a grin as he disconnected his Solasphere.

"I know," Larissa answered quietly.

"I have never seen something so glamorous as the formation of a galaxy. It is simply breath taking," Jendall's voice added to the conversation.

"See, he can be serious when he wants to be," Phelix joked as Jendall looked over at him with a tired but witty smile.

They each reached for another Solasphere and

continued their search.

Like picking up a random rock and skimming it across a pond, LaDon picked a random star, very faint in the distance. This would be his final star for the evening. He did not have high hopes in finding anything useful. If he could just find anything which contained life. Plant life, a microbe, or anything that resembled a living being would make him happy. This star, like all other stars he had visited, matched most. It matched nearly all of the criteria set by himself and the deep space team. The age of the galaxy, the hue, its rate of travel, and magnitude of the sun. They all matched. He moved in closer with each Solasphere, being careful not to get too close and possibly destroying the Solasphere. He noticed the galaxy contained a number celestial bodies just like the ones had seen in previous examples. Before moving closer, he sent in a few Solaspheres before and after the birth of this particular system to see if he noticed anything different. He compared this data to the galaxies he had recorded early in the day. Nothing different.

This lowered his expectations even more as he continued searching the galaxy. He sent the next Solasphere closer to view the planets from the six billion mile mark. This was one of the guidelines given by the deep space team. He watched the Solasphere recording intently. He watched as the timer counted down before porting back to the encasing next to him. Fifty-eight, fifty-nine, and the image went black. Nothing. He sat up in his chair, leaned forward, and clinched his fist until one knuckle finally popped.

"Now you're feeling it," Larissa's voice said softly behind him as she placed her hand on his shoulder.

Somewhat startled by Larissa's sudden presence, LaDon remained with his eyes closed and his face buried in his hands. He could still see the image of the last second of the recording. Like an echo, only visual. Larissa's voice almost allowed him to stay in the moment as his eyes started to see something faint. LaDon started to write it off saying to himself, *This is obviously my desperation to see anything that might resemble life. My mind is probably just playing tricks on me.*

LaDon unclenched his fists and opened his eyes to answer Larissa's question.

Before he could speak, he heard Phelix's voice. "I believe we're all feeling it, my friend."

"I suppose so. I just wonder if we are going about it all wrong. Maybe we're..." LaDon paused midsentence.

LaDon heard his grandfather, Pomph, teaching him a lesson from years ago. *Your mind doesn't play tricks on you, boy. That is, unless you've been wandering through the desert for days without food or water. If something happens in this world, you can bet your rear end something caused it.* With the recollection of such a distant memory, LaDon frantically reconnected his last Solasphere.

"LaDon, are you all right?" Larissa asked as LaDon scrambled for the device.

He flipped the recording to multiple screens

above them and moved the recording forward to the very last frame.

"I need it to be bigger. I need the picture to be bigger. Magnified somehow," LaDon exclaimed as he tried to remember where he saw the tiny speck he thought he saw moments before.

Jendall took the controls and said, "Which quadrant should I enhance?"

"J-4." Phelix's voice held a dramatic, quiet quiver.

"No....it can't be!" Larissa gasped as she rushed to her own terminal to dim the lights.

Larissa returned to LaDon's side as they all gazed at the displays above them. Jendall enhanced the section and the room fell silent.

"It's...it's....blue." LaDon stuttered as each of them stared blankly at the pale blue dot.

Chapter 12

Third Rock From the Sun

"Don't panic, people. Leave that to me," LaDon exclaimed with a worried expression while trying to use humor to diffuse his own anxiety. "I am transmitting the coordinates of the Solasphere to your terminals. Phelix, I want you to monitor the planet's progress going forward in time. Jendall, I want you to monitor going backwards in time. Record every bit of information you can gather. How fast is it traveling around its sun? How far away is it from the sun? Does it have a stable orbit? How much, um, how much...oh, you know what we need."

"We've got you, LaDon. We're on top of it," Jendall shouted as his chair slid gracefully across the floor to his terminal.

"No problem at all, my friend." Phelix hurried back to his terminal.

"Larissa, you move in close to the planet. Try everything you can to get a good look at it without getting too close. If there's any type of intelligence on the planet, they may detect the Solasphere. Maybe try going at it from another angle. Maybe..." LaDon drifted as he shuffled about half a dozen ideas in his head.

"How about I work with Jendall first, since he is viewing the birth of the planet? This way, once I understand how the planet was formed, I can judge the time variable from there. I can get close enough before life begins to form, if there is any, of course," Larissa offered.

"Good thinking. Also, everyone listen. I want to program the Solasphere's sensors to pick up any radio frequencies. If there's intelligence, that will be a good place to start. In the meantime, I need to write up an emergency communication to the Assembly. They must know immediately. This has been their primary concern every time we deliver a daily report," LaDon said as brought up his viewer and began preparing the report.

The team frantically went to work while LaDon fashioned a communication to the Assembly. He took time to write the message as professionally as possible. He knew he came across wordy at times.

They were unsure if the planetary body even supported life. *Should I wait until I have something to report? What if the find is benign?* He hesitated as more and more thoughts rushed through his mind. *This could be one of many planets we find that cannot sustain life.* After debating with himself for about twenty minutes, LaDon decided not to send a report. He wanted to wait until he had something concrete to relay. His thoughts turned to Phelix.

"Phelix, you are coming at this solar system from the end of its life cycle, correct? You are actually the one that will need to be the most careful in your endeavor. If somehow this planet develops intelligent life, we can only assume their technology will reach ours or surpass it. Be extra cautious on how close you approach the planet," LaDon said as he closed the premature message he was about to send.

Phelix did not answer. LaDon saw Larissa look back over her shoulder to see why Phelix did not respond. His mind was focused on the Solasphere recording currently in progress. His facial expression was something unlike any of them had ever seen. He slowly sat up. Obviously the Solasphere recording was over. LaDon walked over to Phelix's station and placed his hand on his shoulder.

"Talk, Phelix. What did you see?" LaDon asked with a firm, supportive tone.

"I lost a few Solaspheres before being witness to the sun going nova. I kept wondering why they stopped returning. After sending a few in at the edge of the solar system, I spotted the problem. The star was expanding. Finally, it went nova," Phelix replied, still looking for his words.

"That's not an issue, Phelix. There are plenty of Solaspheres to spare, and stars go nova all the time," LaDon reassured him.

Phelix looked up at LaDon. It was like he had forgotten his own language and was trying to communicate with LaDon telepathically. By now, Larissa had made her way over to his station. Growing concerned, Jendall disconnected his last Solasphere in order to join them around Phelix's workstation.

Phelix took a slow deep breath and continued,

"It's not the sun going nova that got me. I did my best not to get too close. I calculated the risk of moving in extremely close. I based this on the fact that I could send a Solasphere in just before the sun goes supernova. That way I wouldn't risk running into any intelligence. If I did, they would be dead minutes later. So, I decided to see if I could detect any orbiting satellites around the planet. You know, something similar to our moons. I figured I could simply land there. It does have a moon, so I placed the Solasphere on its surface for a bit of reconnaissance. You know, just to get a better look without getting too close." Phelix's concentration trailed off once more.

"Stay with me, Phelix. Focus," LaDon said sternly to snap Phelix back to reality.

"Oh, sorry. I think I just need to play the recording for you," Phelix said with a blank expression as he reached for the controls.

Phelix faced his terminal and started configuring the console to send the recording to one of the main displays above them. The team watched intently as Phelix prepared the system for playback.

"As you know, when we send the Solaspheres into the new universe, it remains stationary. Normally, if the Solasphere is on auto, it only moves if it notices something of interest or has been preprogrammed to travel in a certain direction. Well, I have all of my Solaspheres set to only move if it picks up anything of interest," Phelix explained as he narrated the recording. "The next thirty seconds will speak for themselves."

LaDon shifted his weight to assure his eye sight was steady. After a few seconds, the Solasphere suddenly shifted its focus as it began to turn. This immediately grabbed the attention of the entire team as each of them shifted their positions while making quick glances at one another. The movement of the Solasphere suggested a possible object. Slowly, the figure of a large, vehicular shaped mass, outlined in gold reflective material, eased into the frame. The Solasphere moved closer to the object and the recording went black.

"What was that?" LaDon asked immediately. "Play it back. Play it back!"

Phelix made the recording start at the exact moment the object appeared in sight of the Solasphere.

"Hold that there," LaDon demanded.

The picture froze. The clear frame revealed the strange object in perfect clarity. They examined the gold, reflective surface with what appeared to be legs used for maneuverability.

"That's not naturally occurring, I can guarantee that," Jendall said in a soft, humorous tone.

"What is it?" asked Larissa rhetorically.

LaDon carefully thought through his plan. He quickly hashed through what they knew up to this point and finally decided that they must visit the planet.

"All right. Here's what we are going to do. If you have any other suggestions, or simply disagree with this plan, speak up. Okay, Phelix, since the sun goes nova shortly after this recording, how about we apply your approach and put a Solasphere on the planet? Sixty seconds worth of readings from a Solasphere can provide so much information, it will take us the rest of the day to analyze it. After we finish analyzing the recording, we will submit our findings to the Assembly," LaDon said in almost one breath.

"Makes sense to me. The last recording got a small bit of information from the planet. It suggests an atmosphere much like our own. We can't be sure though until we get closer," Phelix agreed. "Anyone else? Opinions?"

The rest of the team agreed in unison. All of them were anxious to get a closer look at the planet.

"Great. Since we are all gathered here at Phelix's station, let's send the Solasphere from his terminal," LaDon said as he reached for the closest Solasphere. "Let's place it just inside the planet's atmosphere. We know how fast it's traveling around its sun. We know how fast it's traveling through space. Also, we know how fast it's rotating. Amazing, isn't it? There are slight variations, but it's eerily similar to everything we know about Solaya."

"Yes, it is. We're a little further away from our sun, but our sun has a little more mass than this one. It's almost as if every difference between our two planets has a counter balance which makes it work," Larissa agreed.

"Let us begin," LaDon said sending their minds back to the task at hand.

Phelix input the information to place the

Solasphere just inside the planet's atmosphere. He executed the command. Whoosh! The Solasphere flashed and disappeared. The team waited anxiously for its return. As the enclosure lowered from around the returned Solasphere, the team immediately noticed something different.

"It's not cold," Jendall pointed out the obvious.

"I suppose it wouldn't be. It's remarkable that this device has actually been to another world." Phelix quietly agreed with Jendall. "I love my job."

The team smiled at one another in explorer's delight. Phelix connected the Solasphere to his terminal, which allowed the team to see the recording.

The recording started. Immediately, Phelix dove for the volume control as the sound of the rushing wind was almost deafening. The Solasphere was falling to the ground due to the pull of gravity.

"Oh wow! What I wouldn't give to watch this one through my viewer." Jendall danced around like a child needing a bathroom.

"We will all want to experience this one," LaDon said with a bright smile. "Phelix, turn the camera view toward the ground."

Phelix adjusted the view of the Solasphere to reveal the surface of the planet from high above.

"How far are we from the surface?" Larissa asked.

"We started at sixty from the surface," Phelix answered.

"Here are the readings. The air is a mixture of

nuros and oros. Traces of curos as well," Larissa said as she read the report from Phelix's terminal.

"Curos...but that means..." Jendall stopped and looked toward LaDon.

"Life. I mean, at least life as we know it. How spectacular would that be? Another curos-based life form," LaDon said with wonderment. "We must get one of the Solaspheres to the surface. I have to see it for myself."

"I hope I'm not overstepping my bounds, LaDon, but shouldn't we inform the Assembly?" Larissa asked respectfully.

"You are absolutely right, Larissa. They want to know if the planet is inhabitable by Solayans. First, let's make sure we have this information before we approach them. Prepare the next Solasphere. Get the readings from the previous recording and calculate the rate of descent. Attempt to place the Solasphere just above the surface and see if its hovering capabilities are a match for this planet. I want to know the air mixture at surface level as well. See if you can gather information from the previous recording and locate an open field of some sort. See if you can place it in an area like that. I will prepare our report," LaDon said, confident in his impromptu plan.

The team prepared the Solasphere for transport. LaDon turned his back on Phelix's terminal and walked a few steps away. Behind him, he could hear the whooshing sound of the Solasphere and noticed the flashing of light bouncing off the walls around him. He faded his eyes into his display and accessed the direct information for Alex Cuberly. He watched his display, awaiting a response from Alex. A few second later, his viewer alerted him.

"LaDon! Always good to hear from you," Alex said in a pleasant voice.

"Sir, we have some news for the Assembly," LaDon said with the utmost professionalism.

With a hopeful look and a quick adjustment of his seat, Alex responded, "You found one, didn't you?"

"Yes, sir. We are gathering the final readings now. We have just sent a Solasphere to the surface to assess whether the air mixture is acceptable." LaDon relayed the information as simply as possible.

"Don't let anything see you!" Alex exclaimed with an alarmed shout, startling LaDon a bit.

"Sir, we have been overly cautious on this matter. We are sending the Solasphere in just moments before the sun goes nova. This way, if something were to see the Solasphere, it wouldn't matter." LaDon hoped his explanation would defuse the panic in Alex's eyes.

Alex slowly relaxed back in his seat and took a few deep breaths. He looked as pale as a ghost.

"Very good, LaDon. That is excellent thinking on your team's part. So, what is the reading? Has the Solasphere returned?" The color returned to Alex's face.

LaDon cleared his viewer momentarily and looked toward his team. "Have we been able to ascertain any information on the air mixture?"

The team was huddled around Phelix's

terminal, glued to the viewing screens above them. LaDon looked up to the view screen. Phelix had tied together four display screens, giving a surreal effect to the moment. LaDon's eyes began to process what he saw. A wide field of green and brown plant life blowing in the breeze. A beautiful sunset, unlike anything he had ever seen. A mountain face lining the horizon unlike any formation he had ever encountered. This surprised LaDon, since he considered himself well traveled for a Solayan. He would have loved to immerse himself in the recording. To smell the air and feel the warmth of this new sun all over him. Snapping back to the present, he could hear the irritated voice of Alex Cuberly asking for an update.

"I apologize, sir. My team was playing back the recording from the last Solasphere. I got lost in the moment. The view was breathtaking," LaDon answered.

"And the results?" Alex asked with an understanding tone.

LaDon looked toward his team once more for the results on the test. Larissa slightly averted her eyes toward LaDon and nodded her head gently. LaDon picked up on this gesture and returned his gaze back into his viewer.

"Yes, sir. It's a match."

LaDon watched as Alex quietly celebrated in his seat and beamed with joy. LaDon looked away from his viewer and back to his team. Once more, he looked upon the pleased faces of his team. All of the effort and hard work finally paid off for such dedicated individuals. To know he played a small part in their joy, just the tiniest bit, would make this experience a total success. At that moment LaDon realized how much he had grown to care about this group of new friends and how very proud he was to have this opportunity. An opportunity given to him by the one Solayan he admired just as much as his grandfather. He couldn't wait until his next encounter with Barton Urthorn. They had much to discuss.

"Fantastic news, LaDon! A terrific find in such a perilous time," Alex responded. "I want your team to finalize your research on that Solasphere and report to the Assembly at once. I realize it's an hour early, but I feel with this new information, the meeting may take a little longer than expected."

"Very well, sir," LaDon answered as the display went blank.

LaDon returned to his team. He approached the terminal to find Phelix fully immersed in the last recording. Judging by his face, Phelix was having the time of his life.

He looked toward Jendall and Larissa and said, "Compile as much information as you can over the next ten minutes. We will be taking this information to the Assembly at that time."

"I feel we need more time to analyze the data," Larissa protested.

"I do not believe they are interested in hearing about the last bit of data we gathered. I feel their main concern is that the planet is inhabitable," LaDon said.

Even as LaDon spoke, he realized that Larissa had a point. *Why would this information warrant a visit? There's not much to report until we actually do some research.* Nevertheless, his eyes met Jendall's and Larissa's as he reached down toward Phelix's terminal. They both watched inquisitively as he entered commands in the middle of Phelix's recording.

Access Recording > Copy File > Terminal: LaDon Grafter > Execute.

As LaDon executed the command, a smirk crossed his face. Jendall and Larissa matched his expression when they realized his motive.

"I will be over here if you need me. Just let me know when you have gathered the information and are ready to visit the Assembly." LaDon slipped away toward his terminal.

LaDon took no shame in wanting a little time to himself with the new recording. He approached his terminal, accessed the file, tuned his viewer, leaned back in his chair, and closed his eyes. The recording started.

He recognized the beautiful landscape he saw earlier, but immediately noticed he could not feel anything. He expected wind on his face, the smell of the air, and all the other senses usually picked up by the Solasphere. In seconds, he remembered he had disengaged those senses while sending Solaspheres into space. He stopped the recording, changed the settings, and started the recording again. He had already become familiar enough with his terminal that he did all of this without opening his eyes.

The recording started at zero seconds. The landscape appeared before him once more. The Solasphere was just in reach of tall plant life. He looked outward into the field. It stretched far out in front of his view. He could feel the tips of the plant as it brushed against the Solasphere. He could smell and taste the alien air, which oddly reminded him of home. A sweet, grassy smell mixed with a taste of moisture probably coming from the plant life surrounding him. He soaked in the warmth of the sun sitting high above him just moments away from its own demise. He gazed at the mountain range rising before him. He could see colder spots atop the mountains and green areas around the base. It reminded him of Solaya's mountain ranges in many ways. LaDon let the recording play on a loop.

After the third or fourth replay, LaDon felt a gentle hand running through the back of his hair. He sat up and quickly turned around. He saw Larissa standing there smiling. LaDon shot her a disapproving look and gestured toward Phelix and Jendall, who were both gathering their things getting ready to visit the Assembly after closing up shop.

"Don't worry. They already know, somehow. Did you enjoy your trip to another world?" Larissa asked sweetly.

"Um, yes, I did, but...how do they..." LaDon stuttered, still mentally stuck between worlds.

"We're not stupid, boss. Frequent glances, facial expressions, and body language. They are all empirical evidence at its finest. We're scientists, remember?" Jendall said as he approached LaDon's terminal, followed closely by Phelix.

"Exactly. And don't forget intuition, my friend," Phelix said with a smile as he approached from behind Jendall, and propped an elbow in his shoulder.

Caught like a rat with cheese, LaDon lowered his guard. He stood from his terminal to address them as if admitting his guilt to the entire planet.

"Just know I want to keep things as professional as they have been, please. I hope this doesn't make things uncomfortable for anyone." LaDon looked at each of their faces.

"We're happy about it, honestly. Maybe dinner won't be so awkward next time," Jendall said with a laugh.

"What? Dinner wasn't awkward!" Larissa exclaimed, blushing.

"You two hardly spoke to one another, and you were sitting right next to each other. Jendall and I estimated your first romantic encounter must have occurred the day before or some time just before dinner," Phelix said as if explaining a new theoretical science. "That's why we both left the restaurant so quickly. We wanted to give you two your space."

"Those keen observational skills are going to get us into a lot of trouble or they just may very well save this planet. Let's go see the Assembly." LaDon turned toward the door.

As he exited the doorway, he could hear Jendall from a distance trying to catch up, "So, are we right? We are right, aren't we?"

LaDon kept his pace as he exited the

Observadome, purposefully ignoring Jendall the best he could.

Still where LaDon could hear him, Jendall lowered his voice and said to Phelix, "I knew we were right."

Chapter 13

It Can't Be That Simple

The team made the familiar trek to the Unification Chamber. LaDon thought about how to present this new discovery to the Assembly. Part of his mind lingered on the fact that Jendall and Phelix were obviously aware of his relationship with Larissa. LaDon tried to push this from his mind for the time being and focus on the task at hand. As they down the long, empty hallway toward the Unification Chamber, LaDon had been listening to the conversation his team was having.

"The sky was blue, nearly the same shade as Solaya. The smells, the landscape, and even the tastes in the air were all similar. If I didn't know, I would say we were looking at a recording from our own planet," Phelix explained as the others listened closely. "And the wind whisking around the Solasphere, coupled with the sounds. It was the total experience."

LaDon recalled these same feelings when he played back the recording. Phelix's description was spot on. Listening to Phelix almost took him back there for a brief moment.

They approached the giant, double doors, and LaDon pushed his way into the room. He noticed the same seats lined up in front of the Assembly's table awaiting their arrival. LaDon saw the familiar faces of each member as they made their way to their seats. LaDon's eyes connected with Barton's, who smiled at him proudly. It seemed Alex had already shared the news with him.

"Welcome. We understand from Alex that you have some very important news to bring us this evening. I just want to start out by saying that we know this has been a trying couple of weeks for each of you, and you have put in long hours. You've even given up much of your free time for this project. We want to thank each of you for your efforts. And from what we have heard, the hard work has paid off." Blaine added embellishment to his usual greeting to suit the circumstances.

"Yes sir, we do have news. Also, you're right. This time, the extra work has definitely paid off. Given the readings from the last two Solasphere recordings, we have determined that this planet is indeed inhabitable. There are still many factors we need to address, such as how long of a time period this planet exists in a state that can support life. This will consist of mapping the entire timeline of this planet from beginning to end. This is our plan going forward unless the Assembly sees another approach as more appropriate. Finally, we have reason to believe this planet contains life. Intelligent life, to be exact." LaDon paused for comments from the Assembly.

"Intelligent, you say? What have you found that leads you to this conclusion?" Barton asked quickly.

"Well, sir, we landed a Solasphere on the

planet's orbiting moon to get a better look without getting too close in case the planet did, in fact, contain intelligent life. We did not want to be detected."

"Go on," Barton replied.

"When the Solasphere landed on the orbiting moon, it picked up something on its sensors and moved in to investigate. It found this image." LaDon sent the image to each Assembly member over his display.

The members paused, accessed their viewers to see the image, and turned their eyes back to LaDon at the same time as if they were in sync.

"What is it?" Aleen asked.

"We're not sure, ma'am. We assumed it to be a transport vehicle of some type. We concluded it was definitely not naturally occurring. Some type of intelligent being must have constructed it." LaDon said.

"Did you examine it further? It might have given us an explanation of its origin," Alex answered with a sharp tone, leaning forward in his chair.

"I understand what you are saying, Mr. Cuberly. I did give that a bit of thought. I decided to focus first on ascertaining whether or not the planet was inhabitable rather than searching for intelligent life, so I turned our attention to the planet. I went on a hunch that the object came from the nearby planet rather than some other neighboring galaxy light years away." LaDon explained calmly, hoping not to sound confrontational.

"Makes sense." Alex sat back in his chair with

a pleased expression of understanding. "Good call, Mr. Grafter. I like that method of thinking. Who hired this guy?"

At that moment, LaDon noticed he was a little tense. Once he heard the sounds of jovial laughter from the Assembly, he relaxed. For a moment, he'd thought he was going to have to defend his decision.

Barton stood from his seat and walked around the table to face the team, "Well done, Mr. Grafter, well done. We are most pleased with your team's performance up to this point. Now we have some news for you. As we always try to stay one step ahead, we have been preparing for this moment. We need to know if this planet actually contains an intelligent life form. Obviously, somewhere close by, there is intelligent life. We need to find it. We cannot inhabit this planet at any random point on its timeline because we may alter the history of whatever life form inhabits the planet now. We're not gods. We are merely visitors. We were hoping you would find a planet without intelligent life, but one still suitable to our needs. Then we could simply move there and live out the rest of our lives in peace. Since this is not the case and time is not on our side, at least in this universe, we do not have the luxury to continue searching the cosmos for another planet. We need to gather as much information as possible about this planet, starting from the beginning."

Barton leaned against the table behind him.

"We will start immediately," LaDon said when he was sure Barton was finished speaking.

"Very well. One important thing. We have

synced our meetings with the Lead Representatives to correlate with your evening reports. Don't worry. Those meetings won't be every day. You can guarantee those meetings are going to be packed full of fun, exciting debates. The next one takes place in three days. That gives you two days to gather as much information as you can. The representatives will have a lot of questions, and I want to give them as many answers as possible." Barton said with conviction, shifting his weight back to his feet.

This triggered a change in the atmosphere of the room. the Assembly members rose from their seats and began gathering their things. LaDon's team stood in response.

"That'll be all. Nice work," Blaine said as he motioned his hand toward the door.

LaDon's team made their way eagerly to the exit. Before LaDon could get two steps away from his seat, he felt a heavy hand on his shoulder. He looked back to see Barton. The grip tightened to slow him down. LaDon stopped and turned around to face him. The other Assembly members filed out the back entrance.

"Hold on, LaDon. I would like to speak with you in private if I may."

LaDon heard the big double doors close behind him, which signified his team had also exited the meeting room.

"Sure, Mr. Urthorn," LaDon replied calmly.

"Please, sit with me a moment." Barton ushered LaDon to his original seat, still warm. "And remember, when we are alone, please call me Barton.

Mr. Urthorn makes me feel old."

"Okay. You got it," LaDon said with a small laugh.

For some reason, LaDon felt completely at ease around Barton. He would have told a completely different story a few weeks before.

"LaDon, I understand that Alex had a conversation with you the other day concerning our plan?"

"That's true. He helped me see things from a totally different perspective. He made me realize someone needed to be a few steps ahead at all times. He asked what we would do when we found the right planet. All the time I hadn't even thought that far ahead, until that moment," LaDon answered.

"Exactly. Now, I want you to set aside everything you've learned over the past few days. Go back to that mindset during that conversation. Think when he asked that question." Barton paused, giving LaDon time to jog his memory.

LaDon closed his eyes to concentrate.

"Okay, I'm there," LaDon responded as if participating in a hypnotherapy session.

"Good. Now I want you to think this through. Don't just look at what's happening right now. Look ahead. What's your next question?" Barton asked quietly.

"My next question would be, does it have life?" LaDon responded, keeping his eyes closed.

"Very good. What next?"

"Well, there are two possible answers. Yes or no."

"Very good. For the sake of argument, and based on the object you found on the moon, let's say the answer is yes." Barton seemed to be guiding LaDon toward some pre-ordained thought process.

"All right. If there's life, we cannot interfere. Also, as you said earlier, we can't inhabit the planet before life begins because we might alter the flow of their history. So, we would have to inhabit the planet after intelligent life had ceased," LaDon mused, still unsure where this thought process was taking him.

"Precisely. Now, why would intelligent life suddenly cease to exist?" Barton asked.

"Well, war, famine, drought, or a fate similar to Solaya's. Their own sun's annihilation."

"Exactly. Keep going, LaDon. Reach deep into that historical mind of yours. Now tell me, besides the sun going nova, what would be the cause of an intelligent life form's extinction?" Barton leaned forward as to assist LaDon with his thoughts.

"Their living environment could no longer sustain them. Either the air, water, or food source might have become unusable or no longer available," LaDon said, remembering the fall of many nations recorded in Solaya's history.

"Now, I want you to put everything in one continuous thought," Barton said with a final air.

Once again, LaDon found his face buried in his hands. He journeyed deep into himself. *We need a planet. We have one. It has intelligent life. We can't inhabit it before this intelligence comes into existence because we risk altering their future existence. We can't inhabit it after the intelligent life has ceased to exist, because the living environment would probably prove uninhabitable due to war, famine, or some type of global catastrophe. But that leaves no alternative. We can't live on their barren moon.* LaDon lifted his face from his hands and saw Barton still waiting patiently for an answer.

"Sir, we can't inhabit the planet prior to the intelligent life. We may change something which might stop their initial creation. We cannot inhabit the planet after they are gone, because what destroyed them might possibly destroy us. Things like famine, plague, or lack of some type of resource." LaDon explained, still grasping for his words.

"You are there, LaDon. Push aside everything you know. Push aside everything you believe. Morals, ethics, religious beliefs, remove them all. You must survive, LaDon. You must save your people. This new planet is your only hope. What...do...you...do?" Barton closed the space between them, causing LaDon to slightly lose focus.

LaDon closed his eyes again. This time he stood, removing Barton from his personal space to get some clarity. *Remove my morals? Remove my ethics? What is he saying? Okay, well, if the intelligent life never existed in the first place, that would solve everything. There's no way Barton Urthorn would destroy...*

"Sir, we can't kill them. I mean, we can't stop their creation. It's not right!" LaDon looked up at Barton, startled at his own thought process.

"You're almost here with me, LaDon," Barton exclaimed. "Now, dig deeper. This is why I said remove your morals and ethics. I need you to come to this conclusion on your own."

"But Barton. I'm not killing anyone or anything. My morals will stay put!" LaDon unexpectedly raised his voice to the one Solayan he respected more than anyone else.

"No, no, no. We're not going to kill them, son. That's our entire reason for not inhabiting the planet in its early stages. We want to avoid stopping their evolution by accident. But, I had to get you to this point so you could take the next step. Think, LaDon, you're almost there," Barton said as he closed the distance between them once again.

LaDon turned away from Barton, mixed emotions furiously tormenting his mind. *If the answer is not in their creation, does it exist within their destruction? How or why do they cease to exist? In this scenario we are playing, why does this life form cease to exist?*

"In this purely theoretical story we are playing out, how do they die? Why do they cease to exist?" LaDon asked with closed eyes when he realized he didn't ask his question aloud the first time.

"All right, I'll play along. Nuclear fallout. A World Ender, if you will." Barton replied, sharpening his gaze on LaDon.

"Then the planet is uninhabitable once they are destroyed, even for us. We may not be able to repair the atmosphere, and we don't have the time to figure that out," LaDon replied with equal swiftness.

Barton raised his eyebrows and smiled, "So, what do we do, LaDon?"

"We can't let the planet become uninhabitable...", LaDon paused midsentence as his stomach turned over on itself.

Time stood still for a moment. LaDon started to feel dizzy as unfamiliar emotions began coursing through his brain. The realization of what he was about to say was almost too much for him to handle. The one thing they had sworn to avoid was the one thing that made the entire scenario make sense.

LaDon looked to Barton, whose expression had changed to that of acceptance and relief. LaDon walked over to the edge of the table. He placed both hands on the table and looked down at its smooth, gloss surface. He could feel the condensation forming on the palms of his hands. He took a deep breath, collected himself, found his inner strength, and turned to Barton.

"We have to make contact with them, don't we? It's the only way," LaDon said as the realization of his words slowly sank into the fiber of his being.

"I knew you had it in you. That imagination of yours is not to be trifled with. You are your grandfather through and through, son." Barton said slowly and with conviction.

"But we don't know if they have ceased to exist," LaDon said.

"Even if they haven't. What if they lived up to the very moment their sun caused their demise, just like we will? We need a planet, and we don't have a lot of time to go searching for another one. We're not murderers, you know that. This leaves us with no alternative. We must talk to them, somehow. And before we can do this, we need to know them, from start to finish. We need to know everything about them. We need to know them more than we know ourselves. This is where your expertise comes in, my friend," Barton explained as he preached to LaDon in the same political manner he used during his political updates.

"My expertise?" LaDon's face turned with a question.

"Yes. We must build a complete timeline of this planet. Start from the beginning. Map it out. We want to see it all. Once we understand their languages and the limits of their technology, we will need to know the most opportune time for us to make contact. If we walk in on them in their early stages of development and tell them we are from another world, let alone another universe, the consequences could be disastrous. They must be prepared for our arrival. Imagine if three months ago, someone came to the Assembly and said they were from another universe. We would probably have them committed to a mental institution or simply thrown in a psychiatric penitentiary. We must know the best time to make them aware of our existence. And then, of course, request inhabitance for our entire planet's population until we can find another planet to call home. Just making the space we need would require time and resources. Our Nuweeyan architects will be in for a surprise." Barton slowed his speech as he came to the end of his thought. "Your expertise on history makes you the best man for the job."

"I know what we have to do." LaDon almost made it sound simple. "We will get started in the morning. I believe I need to go home and rest."

"Yes, as do I." Barton smiled as they both stood from their seats. "I enjoy our little chats, don't you?"

"When we have a *little* chat, I'll give you my opinion on it," LaDon quipped as they journeyed toward the double doors.

"LaDon, I want you to know one thing before you go." Barton stopped LaDon just before opening the door. "I chose you for this job for a reason. You are the most noted historian on the planet. As you and I have discussed, your grandfather Pomph and I worked together on many occasions. I respected him greatly, just as I am growing to respect you. You are more like him than you realize. You're bright, analytical, and focused on whatever task is laid before you. I want you to know, on a personal level, I am here if you need me. Not in a formal Assembly meeting and not over your viewer. I mean, I'm here. We're asking a lot from you, LaDon. We realize this. I realize this. Please, do not hesitate to come to me if you feel lost, confused, or simply need someone to hash out a plan."

"Thank you, Barton, sincerely. This means more than you realize. I've always admired the Assembly from afar, through the eyes of the Solaspheres. Now, working in close quarters, I am beginning to feel the same way as you. And after this talk, I believe I just might take you up on that offer over the next few day pass," LaDon replied with a sincere heart as he reached for Barton's hand.

Barton grasped LaDon's hand firmly and made deliberate eye contact to convey his feelings even more than just a handshake.

"Well then, I look forward to our next meeting, LaDon," Barton replied with a heartfelt expression.

LaDon realized his personal meetings with Barton were never going to be easy. It seemed their meetings were always to be filled with emotional highs and lows mixed with deep, philosophical questions. But this time was different. Yes, there were still emotional roller coasters and deep thinking, but this meeting ended on a completely different note. He now saw Barton as more than just a figure of admiration. He saw him more than just the leader of the mighty Assembly. He considered Barton a friend. A good friend. His wisdom and poise during such a perilous time spoke volumes to LaDon.

Truly, he is a great man.

Chapter 14

They're Alive

The next morning, LaDon woke early. His head was filled with activities from the day before. Also, he couldn't stop thinking about the talk he had with Barton. The man he had once admired now seemed more like an average Solayan rather than a figurehead that sat high atop their throne. He realized Barton had doubts and fears just like every other person. Still, LaDon admired Barton more than anyone he had encountered in his lifetime. Maybe this feeling was from years of admiration as a child. Maybe there was something else. LaDon couldn't quite put his finger on it. He felt one day he may be able to solve this puzzle. Until then, there was work to be done.

Taking advantage of rising early, he prepared himself an elaborate breakfast before heading out. After his meal, he stepped out of his house toward his transport. Just as he reached his vehicle, he received a communication over his viewer. LaDon stepped inside his transport and responded to the transmission.

"Yes?" LaDon said without reading the text showing the incoming caller.

"Hey, you. Up early this morning, I see." Larissa's voice sweetly echoed through the viewer.

Such a pleasant voice and lovely face on such

an exciting day. He realized they hadn't spent much time together these last few days and was pleased to hear from her.

"Hello, my sweet. How are you this morning?" LaDon said in a cheery voice.

"I'm doing great. I've been missing us lately, though. We should do something tonight," Larissa said with a pleading tone.

LaDon could tell from her voice that she had something more to say than the normal morning pleasantries.

"That sounds like a great plan. It sounds as if you have something else on your mind. Is everything all right? I see you are already at the dome." LaDon hoped his intuition was wrong.

"Well, I suppose I can't say anything is wrong, in particular..." Larissa's voice faded into the end of a sentence.

"But...?" LaDon prompted, trying to pull the information from her.

"Okay. I have slipped outside the Observadome for a moment. LaDon, the Assembly has set up shop in some of the seats around the edge of the Observadome. They have removed some chairs and put in a small work area. They're all here. Almost like they're watching a sporting event. Did you know anything about this?" LaDon could tell that Larissa was keeping her voice low to avoid causing suspicion.

"No. This is all news to me. Have they addressed the team? Have they said anything to you?" LaDon asked, even though she would have told him by now if she knew.

"Of course not. Why do you think I am calling you?"

"I know. I know. I was just thinking out loud. Well, it seems I didn't wake up as early as I thought, otherwise, something is up. Have Jendall or Phelix arrived?" LaDon asked.

"Phelix has. Jendall will probably arrive soon. How far away are you?" A hint of concern entered her voice. "I'm kind of scared."

"Don't be scared. I had a long talk with Barton last night. Everything will be fine," LaDon said with empty confidence.

"Last night? You said you don't know what's going on. If you talked to him last night, he must have said something to you. What did he tell you?" Larissa asked .

"I promise you. He didn't mention a word about what is happening right now. Maybe he didn't feel it was necessary. Maybe he wanted me to keep a clear head and focus on the task at hand." LaDon said aloud with confidence while his mind was riddled with doubt.

"Well, I don't like this, LaDon. This worries me. Are we doing something wrong? Are we not moving fast enough?" Larissa's voice rose as her worry increased.

"Look, Larissa. Please calm down. Don't worry. I will straighten everything out when I arrive. I am getting closer by the minute. Think of it this way. When have you known the Assembly to make any move without due cause or logical reason?" LaDon paused hoping to calm Larissa.

"Yeah, maybe. Just get here quick." Larissa said, sounding as if she wanted to end the call.

"I'll be there soon enough. Try to think of a place we can go this evening. Maybe that will get your mind off worrying long enough for me to get there." LaDon continued to attempt at derailing her thought process.

Larissa sighed. "Okay, all right. I know what you're trying to do. I'll try and calm down. Just hurry. I miss you," Larissa said as her breathing somewhat slowed.

"I miss you too. See you soon. Goodbye," LaDon said as he disconnected the session.

What could they possibly be doing? Setting up a work area surrounding our terminals? Are they taking over the project? And why didn't Barton tell me this yesterday? He had all the time he needed to tell me their plan. But when do they ever tell anyone their plans? These thoughts and many more raced through LaDon's mind as he approached the parking area of Nalkalin.

He stepped briskly toward the Observadome, determined to get to the bottom of this. Just as the lift arrived at the Observadome, LaDon stepped through the main doors and assessed the battlefield. It was just as Larissa had explained. Surrounding the Observadome floor, directly behind their terminals, Solayans moved equipment into an area that was once chairs that laced the outer border of the Observadome. A long table similar to the one in the meeting room was stretched where chairs used to exist. A couple of terminals were already installed at each end of the table. The setup matched the Unification Chamber. Those terminals were obviously for Blaine and Alex.

LaDon approached one of the busy figures and asked, "What exactly is going on?"

"We're not sure, sir. We've just been given these specifications on how to install this equipment. All I know is we received the order from the Assembly this morning," the worker stated as he continued his task.

LaDon's focus left the worker. *This morning? Why didn't Barton say anything about it last night?*

LaDon felt a presence behind him. He turned to see Larissa standing there, already wanting answers.

"So? What's going on?" Larissa said with a slightly more demanding tone than he heard just moments ago.

Obviously, the short time between their communication and now had caused a small lapse in her relaxation technique. Possibly she was done picking a place to eat just in time to continue her fretting. LaDon placed his hands on her shoulders and spoke calmingly.

"Larissa, look at me. I just arrived. I am going to sit my things down over here by my terminal. Then I will find the Assembly and get some answers. The worker over there told me they just got the designs this morning. Maybe that's why Barton said nothing to me last night. They are quite secretive sometimes. You know this. Stay close by me and we will get to the bottom of this together." LaDon watched his words have a calming effect on the frazzled but beautiful Vaknoreeyan that stood before him.

He walked toward his terminal with a calmer Larissa in tow. As he placed his things beside his terminal, he saw Phelix approaching with Jendall not far behind. Both scientists started talking at once, but it came out in a jumbled mess.

"My friends, listen, I am about to attempt to locate the Assembly and find out just what..." LaDon started to explain.

Just then they each heard the main entry doors swing open. They turned to see all four members of the Assembly entering the room just as the workers were starting to clear out. LaDon glanced over his shoulder to see the work area was complete. *Their timing is impeccable,* LaDon thought as he stepped forward out of the pack to greet them.

"Welcome. We're very glad to see you all. May we inquire about the sudden change?" LaDon asked with plausible inquisitiveness.

Barton stepped around to be face to face with LaDon.

"I know this is sudden. We do ask for your forgiveness, but we only came to this conclusion late last night. LaDon, after our discussion, I called the Assembly together for an impromptu meeting. We discussed each of your teams' findings in detail. The recording of the open field. Also, the air mixture being perfectly compatible with Solayan chemistry. We decided this is the planet we need to survive.

With your discovery of the object on the moon, we believe this to be proof of intelligent life. We also believe this is the last leg in our journey to save our world. Your team will bring about the result in this endeavor. This is Solaya's last hope of survival. We feel this is where we belong." Barton paused to take a breath and ascertain the mood of the room before continuing. "We have all the information we need. We must get to work on mapping the timeline of this planet. We need undeniable truth that this world contains intelligent life at some point along its existence. If it doesn't, we will simply inhabit the planet. If it does contain an advanced form of life, we must not interfere with the development of that species. The entire Assembly agrees that would be unethical. If life exists, we must figure out a way to communicate with them that does not interrupt the flow of their history and still allow a way to inform them of our existence."

"That almost sounds impossible, sir." Phelix snapped during the small break of silence.

"Yes, Phelix. This will be tough, but it's our only choice. We have thought this through. We must hope they advance their scientific knowledge far enough to believe at some point that we come from a different universe than theirs. From there, we can request to exist peacefully with them until we find a place of our own. If we pick the wrong time during their evolution, they may think us gods or witches. This outcome would simply result in violence and disbelief." Barton pursed his lips, took a breath, and began to talk with his hands. "We must think through this logically with multiple avenues and opinions as we make these decisions. These decisions are going to ultimately shape the fate of our nation. Therefore, we will be stationed here throughout the duration of this project. On a personal note, please understand, this is in no way a slight on your progress up to this point. We feel we need to be in close quarters because we feel this planet needs every available resource to accomplish such a task."

"As always sir, we will do our best." Jendall spoke up in response to Barton's explanation.

Barton slowed his speech which added a little drama to his next point, "Good, very good. And we're behind you. Each one of you. If we succeed, we succeed together. If we fail, we will fail together."

"Let's get to work then!" Alex Cuberly's excited voice caterwauled from the Assembly's table across the room.

"Very well then." LaDon turned to his team. "Are we ready?"

LaDon looked at each member of his team. Their faces beamed with determination after the mind altering speech from Barton Urthorn. His words carried such weight, who wouldn't want to follow such an accomplished leader? LaDon's eyes rested upon Larissa to find her face just as excited after knowing the full story behind the Assembly's presence in the Observadome. This gave LaDon the extra bit of confidence he needed to execute the plan he formulated the night before.

"All right, Jendall and Larissa, start from the

formation of the planet and work your way forward. Phelix, start from the explosion of the sun and work backwards. Again, let's make sure we are discreet with the Solaspheres. We don't want...", but LaDon was interrupted by a voice from the Assembly as it began making its way down to the main floor.

"LaDon? LaDon? I have something which was designed by one of our teams. The credit has already been given its proper due. Those Kalloneeyans are at it again," Alex explained as he made his way to the team's huddle.

"What do you have, Alex?" LaDon looked on inquisitively.

"Look at your terminal. Now look inside the Solasphere encasing," Alex said with a smile.

"I see nothing inside the encasing, sir." LaDon gazed at the empty enclosure.

Out of nowhere, Phelix yelped and approached LaDon's terminal. "No! They didn't figure it out, did they?"

"Just enter some coordinates," Alex said smugly.

"But there is no Solasphere attached," Jendall explained as he walked closer to the terminal.

"Jendall, the Light Breaker," Phelix said as his face looked at Jendall with amazement.

"Nooooo! But how did they get around...? I mean how did they compensate for...?" Jendall stammered, at a rare loss for words.

"Is it really in there?" Phelix turned toward Alex for an explanation.

"Yes. It is," said Alex with a bright smile.

"Is what in where? What are you two babbling about?" Larissa asked as she approached the terminal.

LaDon walked toward the enclosure and flipped a few switches to lower the mystery enclosure. As he reached toward nothingness, LaDon's fingers buckled as they collided with thin air.

"Ouch!" LaDon exclaimed, jerking his hand back.

He reached out again into the thin air, slowly. He made contact with the empty space. He began tracing his fingers around the border of the blank area. He closed his eyes in order to encourage his sense of touch. He began to feel ridges, smooth metallic surfaces, and even multiple glass lenses. He immediately realized he was touching a Solasphere.

"Is there really something there, LaDon?" Jendall's voice held disbelief.

"This is a Solasphere I am touching. I guarantee it. I have handled enough of these, I would stake my life on it. This is a Solasphere, but I can't see it.", LaDon said still groping the Solasphere relentlessly.

"We will be equipping every Solasphere we have with this technology. We can send these devices whenever we want to record without being detected. I think it's time to begin, don't you?" Alex asked with a smile and a nod.

"Yes! Let's continue as planned. Now we don't need worry about being so cautious. Let's be mindful not to bump into anything, though," LaDon addressed his team as they prepared their terminals.

The team learned how to use the new, cloaked Solaspheres. They began sending probe after probe, storing information according to time and location. Jendall and Larissa moved forward on the timeline. LaDon teamed up with Phelix to work backwards in time. Both teams closely watched for any type of intelligent life.

The entire team was engulfed in their current Solasphere recording when a sudden crash came from Jendall's terminal. Everyone immediately disconnected from their terminals and rushed over to aid the fallen Jendall. Alex rushed down from the assembly's area. All other Assembly members quickly stood from their seats to get a view of Jendall lying on the floor. Oddly enough, Jendall was still engaged in his recording. Everyone could see his face. His mouth was wide, bright, and full of life. In a panic, Phelix reached for Jendall's terminal to disengage the recording to bring Jendall back to reality.

The sudden disconnect was a slight shock to Jendall and he immediately started bellowing, "No! Turn it back on! Start it again! This is absolutely amazing!"

Jendall lunged for his terminal and started the recording over again.

"What? What do you see, Jendall?" LaDon asked with a crazy laugh, still shaken by the large crash.

"Let's find out, shall we?" Larissa reached for his terminal and copied his recording over to her own.

The audience moved to Larissa's terminal. She combined a few monitors high above for everyone to see as she started the recording.

As the recording started, they could hear Jendall's voice coming from his terminal, "Here, let me send you a copy. Oh, wait, you're already watching it. Well, you see, I decided since we were cloaked, I would send the Solasphere far from the initial creation of the planet just to see..."

But there was no one listening to Jendall's rambling. Everyone was glued to the monitors above. The booms rumbled the sound system within the Solasphere as the giant, colossal monster glided along the display. The Solasphere obviously saw the animal as a point of interest and followed closely behind. It was apparent now why Jendall fell from his seat. The booms remained intense as the Solasphere trailed the creature. After the initial shock of the discovery, the team finished watching the recording and regained their focus.

"That creature is proof that something other than plant life exists on this planet. This is great news, my friends. Great news! Let's continue the search. And Jendall, try not to scare us to death next time. It's just a recording. It won't bite," Blaine explained as nervous laughter allowed everyone to shake off the jitters caused by the event.

As the team drifted back into their research, Phelix called LaDon over to his terminal.

"LaDon, I'm getting something strange on the sensors of this last Solasphere. Readings of highly concentrated curos in the atmosphere. It almost seems to be encompassing the planet," Phelix explained as he showed LaDon the new readings.

"What could cause this buildup?" LaDon wondered aloud.

"Massive burning of materials caused by intense heat. Very intense heat, to be exact." Alex's voice rang out as he wandered back to the Assembly table from the encounter with Jendall's monster.

As everyone's attention went back to their work, Phelix quietly whispered to LaDon, "Boss? Watch the last five seconds. I have stayed in this field to get readings before venturing around the planet, but I noticed something this last go around. I just happened to look in all directions. Look out at the horizon at the base of the mountains."

LaDon took a long look as his eyes adjusted to the monitor. He saw the mountain range. Next, he looked along the base of the mountain range where it meets the ground. Suddenly, his eyes fixated on the spot Phelix was pointing toward. His eyes began to make out a small structure. Looking closer, LaDon realized this was not a naturally occurring structure. This structure was too complex to be created by a simplistic life form.

"Send in another one! Put it down right next to the structure, Phelix!" LaDon exclaimed, breaking their secretive silence.

This alerted the Assembly, which was LaDon's ulterior motive. They all remained in their seats although their attention was fixed on Phelix's terminal. Jendall and Larissa disconnected their sessions and gathered around Phelix's terminal.

As the Solasphere was sent and returned, LaDon noticed that Barton had made his way down to the floor and positioned himself for a better view. He placed his hand on LaDon's shoulder for support as the monitor was activated. The Solasphere began to pan around as it examined the structure. It appeared much like the early designs of Nuweeyan technology. Entryways containing mechanical pads. Much like the pads found in lifts around Solaya. The Solasphere zoomed in on the strange symbols found printed on the pads lining the entry way.

As the team leaned in to get a closer look at the monitor, Barton's voice broke the silence.

"They're alive."

Chapter 15

History Repeats Itself

Barton stopped pacing as he faced the team, "All right. There is definitely a species at work here. I want to know everything there is to know. I want our goals to be as follows. First, what life form made these markings? Second, did they originate from this planet? Finally, if they did originate from this planet, how did they evolve? How does their existence end, or does it end at all? To know this information, we will need a full map of their timeline. LaDon, I want you responsible for what part of the timeline should be explored first. Remember our discussion about the Assembly thinking ahead? Well, we've caught up with each other. That's another reason why we are in here. the Assembly knew if we found life, we would need to make contact. We also knew, to contact the life form, we had to know what they know. We must make them understand the fact that we come from a separate universe altogether. Am I making sense here? Talk to me."

"Perfectly clear," Jendall answered.

"Clear," chorused the rest of the team.

LaDon addressed his team. "Jendall and Larissa, move your time variable closer to this point in time. I doubt the monster you encountered has any chance of making those marks. Funny, that creature resembled the Balosors skeletons on display in the museum. Doesn't matter. Phelix, you and I will continue our quest backward. First, I want to investigate this facility. Let's see if we can get a Solasphere inside the building without having to penetrate the entryway. Everyone know their task?" The team was already quick on their toes and executing his commands. "Very well, then."

The team began their search. They began sending Solasphere after Solasphere into the new world, mapping every inch of the planet's surface. Discoveries began to multiply like wildfire. Buildings, flying machines, roadways, and varied life forms all flooded the Solaspheres' storage banks.

LaDon was most perplexed by the physical makeup of the life forms. Standing erect, walking and talking. They had two eyes, one nose, one mouth, two arms, and two legs. Other than the rounded off chin and a variety of eye colors, they could all pass for Solayans. LaDon had never seen an eye color other than blue. This intrigued him. He began programming his Solaspheres to get close enough to these beings to see their eyes. LaDon took a liking to green eyes. *Who would have thought? Green eyes. They almost look possessed.* LaDon laughed to himself.

As for skin tone, there were a variety of colors. He saw some with very dark skin while others were much lighter shades of peach. There were plenty of colors in between those shades as well. This was also interesting since all Solayans had the same tone of light brown.

Their hair even seemed to follow this trend.

Such diverse colors such as reddish orange, whites, yellows, browns, and blacks, leaving nothing left for the imagination. *How can one even keep up with it all?* LaDon said to himself.

Aside from the visual, they heard strange sounds coming from the lifeforms. The speech pattern was unlike anything they had ever encountered. Even more confusing, the speech patterns differed as the visual attributes differed.

"Linguists. We need linguists in here now, sir," LaDon told the Assembly.

"Good call. Right away, LaDon." Barton motioned to Blaine.

"Larissa, Jendall, and Phelix, please gather as much audio as you can for the linguists when they arrive. Who knows, maybe the Unifier can make some sense of it," LaDon commanded as his team shifted their focus.

Blaine activated his viewer and began talking immediately as he stepped away from the table. LaDon watched as Blaine exited the room.

LaDon had a small flashback to a particular Solasphere recording. It was the recording where the Unifier got its name. It was during the Forgotten Wars when the Vaknoreeyans and the Kalloneeyans were having much difficulty due to their language barrier. It seemed like all hope was lost in gaining a peaceful connection between the two races. This was definitely an issue until the invention of the Unifier by the Delnokeeyans. The Delnokeeyans saw that without peace between the Vaknoreeyans and the Kalloneeyans, the war would never end. Their land fell directly between the two nations.

It took almost a decade to perfect the Unifier. Once it was prepared, the Delnokeeyan leader called a meeting on neutral ground between the two warring factions. The moment the Vaknoreeyan leader spoke and the Kalloneeyan leader heard their words for the first time, it was like their feud ended at the meeting table. Tensions released, cease fires were executed, and even sighs of relief were shared. Both sides realized they were fighting for very similar reasons. Reason concerning family, friends, and even their own pride. Peace between the two nations still took years to settle, but that meeting started a trend of peace that eventually developed into present day Solaya. The Unifier earned its name that day. Everyone knew the story of the Unifier. LaDon felt this device may relive its glory days once again.

Days, weeks, and months passed as the team gathered data. Just as LaDon predicted, the Unifier took each language pattern found on the planet and decoded them.

Moment by moment, era by era, century by century, LaDon's team began to learn many things about these beings. They learned that the beings referred to the planet as Earth. They learned there was only one race of sentient beings, the human race. The team learned much about their diverse cultures, their history, and their ways of life. They watched the species evolve throughout the passage of time. They watched them grow from humans who could barely communicate, to humans living in caves, huts, and even building small homes of wood, stone, and eventually iron and other metals. Their technology advanced up to the point of space travel.

LaDon found himself distracted after finding articles from Earth discussing deep space travel. *Wow, they even get into deep space travel. It doesn't look like they have ventured into any type of warp technology at this point. It looks here as if they tried to reach the red planet just one orbit path over from their own. I wonder what's over there.* LaDon's curiosity got the better of him.

He started sending Solaspheres over to the red planet the people of Earth called Mars. LaDon liked "red planet" much better. He knew that Phelix was continuing to map the more advanced civilization era of the timeline, so he felt the research was in good hands for a few hours. After successfully placing a Solasphere on the red planet, LaDon began by letting the Solasphere stay for longer periods of time and wander the surface of the planet. The planet seemed barren, with no signs of life. The atmosphere was not capable of supporting Solayan life. He referred back to the article. It mentioned a pod launched toward the red planet. LaDon took some educated guesses as to the time and place and let his console do the math. Using the landing date and time mentioned in the article, he did his best to pinpoint the location of the planet at the time the pod landed. Just as his terminal predicted, he found it. *Hah! I found it. Little victories.* He entered the commands for the next Solasphere to get a little closer to the object mentioned in the article. Just then, he felt a presence behind him. He turned to see Phelix standing there.

"Sir, I have completed our research of the end of the timeline," said Phelix in a somber tone almost close to tears. "It doesn't look good, LaDon."

"Why?" LaDon asked while a feeling a dread crept across his face.

"They're all dead. Do you remember all the black carbo we found a few months ago? It was right around the time where they lose their sun. Now, remember we thought it was their sun's destruction that caused their demise?" Phelix paused and looked at LaDon for a response.

"Yes, I remember," LaDon answered to encourage Phelix to spit it out.

"Well, it's not the sun that causes them to meet their end. They are extinct way before their sun explodes. They destroy themselves, LaDon. With their own technology, they obliterate themselves," Phelix said in disbelief. "After I learned this, I sent a Solasphere high above the planet to watch for myself. It was like I had a front row seat to the reenactment of the Forgotten Wars. Immediately, I saw it. They launched a World Ender. That's when I lost it. The people of Earth call them nuclear bombs. They launched one, LaDon. They actually had the audacity and the ignorance to launch one! Of course the rest of the factions followed suit, which decimated the entire planet!" Phelix finished his sentence with tightened fists as his voice cracked and his eyes welled up with tears.

the Assembly members and the rest of the team heard Phelix's passionate words and made their way over to LaDon's workstation to hear the rest of the news. Aleen came up close behind Phelix and placed a caring arm gently around him.

Phelix attempted to contain himself but his words continued to pour from the spigot jammed directly into his heart, "There's so much more to live for. So much more to life than pointless fighting over who gets what piece of rock. It was like watching waves crashing into each other. Bomb after bomb, explosion after explosion. It's pure lunacy, I tell you, lunacy!"

Phelix let out one more vocal sigh as his passionate plea for life came to a close. The group stood silent for a moment as they contemplated the implications of Phelix's words. Depressed but reverent looks found their way to each of their faces. It was obvious all the hearts in the room shared in Phelix's anguish. Memories of their own world flashed through the minds of the entire room. LaDon looked up at the group through blurry eyes and could see the faces of the people he had grown to admire and love over the last few months. He saw Larissa's face. He could see where tears once lived, leaving shimmering tracks as a reminder of their existence. No other voice at that moment could break such silence than the voice of Barton Urthorn.

In a soft, consistent cadence, Barton spoke, "They sound much like Solaya. Back in a time when our planet was as its worst. Fighting for power and pride. Fighting for land and resources. All in a feeble attempt to gain control of something that will always elude them. There is no winning when you must kill.

All life, life on Solaya, even life on Earth, is precious. You know, I have always considered us lucky as a species. Our timeline unfolded in such a manner that technology saved us. This shows it could have very well been our end. Let's break for a few minutes to gather ourselves. Spend this time in thought."

The team agreed. They returned to their terminals and the Assembly returned to their seats.

LaDon felt the emptiness in the air increase as each person retreated back to their area. One person remained. It is the one person he hoped would remain near him. Larissa stepped in front of LaDon. He rested his head in her hands. She stood there in silence for a moment as she attempted to comfort him. LaDon began to contemplate his own mortality, realizing that everyone else was probably doing the same. He thought how good his life had been up to this point. No real hardships, no fear of war or famine. He was even granted a wonderful childhood full of stories and laughter.

Now he was here. Larissa, a beautiful Vaknoreeyan who loved him dearly, a team he admired more each day, and the respect of the entire Assembly. Not to mention, the friendship that he had developed with Barton Urthorn.

At one point, Barton was so high on a pedestal, LaDon could never see himself reaching such heights. It wasn't until starting this venture that LaDon finally saw Barton as another Solayan. Nevertheless, a great Solayan. LaDon felt as if Barton was a type of father figure, guiding him through this journey.

LaDon needed this time to reflect. He saw his wonderful life and the wonderful people he had the pleasure of knowing. He had a chance to save them all, and save the world in the process.

Yes...save...the...world... Why can't we save the world? While we are saving Solaya, why can't we save them? A new sense of hope stirred in his heart. After what was only five minutes or so, LaDon abruptly lifted his head from the cradle of Larissa's hands and stood from his seat. Larissa stepped back to allow him room to move. Her face was perplexed trying to figure out what just entered LaDon's mind. LaDon saw the assembly members sitting at their table. Barton was leaned forward, looking at a blank area on the table and Aleen gently massaged his back. LaDon took a few steps toward them and felt Larissa's hands holding his arm.

"Are you all right?" Larissa whispered softly.

LaDon smiled at her with reassurance, resisted her grip, and turned his attention back toward the Assembly's table. He took another step closer in their direction.

"Sir? If I may suggest something." LaDon paused, awaiting their attention.

Barton slowly raised his eyes to meet LaDon's. The other Assembly members fixed their gaze on LaDon as well.

"Always, son. What's on your mind?" Barton answered with a small grin.

"From the looks of the timeline, this volley of terror not only renders the planet uninhabitable, but it remains this way up until the explosion of the sun.

So, obviously their technology never reaches a state where interspatial travel is even possible. Our logic is correct. The planet is never inhabitable at a time where we don't run the risk of interrupting their flow of history; therefore, we must contact them. I mean, how else would they ever believe us?" LaDon paused for his words to have time to sink in.

"We've been here before, but go on. What are you getting at?" Barton answered.

"We already see how the planet's history unfolds much like our own, right?" LaDon looked around the room for affirmation.

All heads in the room nodded slowly as everyone in the room was listened to LaDon's words.

"Who's to say we can't fix it? We could fix their history. Repair it somehow. Keep them from destroying each other. We could save their world while at the same time saving our own. We need them to survive, and, from my perspective, they need us to survive. We can stop that bomb from launching. Heck, we can even stop the reasons for having the bomb in the first place. With time as our friend, the possibilities are endless." LaDon's voice grew slowly more intense after the end of each sentence.

"But isn't that interfering with their timeline? Haven't we sworn not to do this?, Phelix said from his terminal.

"They die, Phelix. And when they die, we die along with them. What other alternative is there? We can't find another planet. The time it would take to find another planet and know everything there is to know would be too great of a risk. We must act," LaDon urged.

LaDon shifted his eyes back to the Assembly. Barton and Aleen were having a small conversation. Blaine looked in LaDon's direction, rubbing his hand across his beard in thought. LaDon then saw Alex making his way down to him.

"Just how do you intend to make these changes, LaDon?" Alex asked.

"Well, I don't know exactly. We could program a Solasphere to enter the area where the World Ender was launched and have it emit a pulse which could knock out the system." LaDon thought quickly off the cuff.

"And then what? Remap the timeline from that point and see what happens?" Alex asked, playing devil's advocate.

"Well, yes, that makes sense." LaDon answered.

"And then what do we do? Wait until the next poor sap presses the launch button and then stop him?" Alex said more sternly than before. "We have no idea of knowing what might happen next."

This caused LaDon to stop and think. He knew he hadn't thought this through and his words were premature. *Maybe they shouldn't have been spoken at all. Maybe that's why Alex has reacted so suddenly and harshly,* LaDon thought, hashing his thoughts out in his head. LaDon tried to ignore the hovering presence of Alex Cuberly. *What if we change something? What is the effect? Can we go back and undo the change? How can we even test this? We can't. We only have one shot at this. He's right. How do we know what will happen next?* Suddenly another brilliant lesson from his grandfather, mixed with words spoken by Barton just moments ago, suddenly collided. The answer came rushing into LaDon's mind like a dam breaking on a small village.

"Mr. Urthorn, you said something a moment ago." LaDon's voice broke the silence.

Barton had been looking at LaDon for some time now, he realized.

"And what's that?"

"You noted we all see how Earth's timeline is similar to that of Solaya. All the way up to times much like the Forgotten Wars." LaDon paused but gathered his words quick enough to continue without warranting a response. "Their world's history very similar to our own. Races struggling for power. Wars breaking out over resources. There's something my grandfather taught me that lives within the root of everything I believe. History repeats itself. History is one of the few predictable things we have in this life. I say we apply this logic here."

LaDon smiled while tasting the sweetness of the memory.

"I believe I see where you are going with this. Tell me, what exactly are you suggesting?" Barton shifted his weight in his seat.

"I say we overlay our history on top of Earth's timeline to match our own. Moments in time where power shifts. Events where wars end simply by the advancement of technology. What saved us so many years ago might save them." LaDon looked around the room at the faces slowly starting to understand his thinking.

"What about the Treaty of Balgeron? The Declaration of Kalloneeyan Arms? Moments like this were the actions of Solayans, not the advancements of technology," Larissa mused as she tried to make the theory into a reality. "Sure, we could introduce technologies, no doubt. But those technologies wouldn't even be possible without struggles put to rest and alliances formed. All of this occurred when peace talks and language barriers were crossed. We would have to get someone from Earth to play these roles in their history, but that would mean communicating with them prematurely. We can't do that. It would need to be someone that knows the history. Not only someone that knows our history, but knows how those events played out. But that would mean, somehow, someone from Solaya would need to actually go there. Live there at different points in time. A Solayan would have to go...I mean...wait...is that even possible? No, that's not possible, right, Phelix? Jendall? We can't send..."

Jendall and Phelix looked at Larissa with perplexed expressions, when suddenly Barton stood from his seat.

"Jendall? Phelix?" Barton's firm voice resonated through the void of the Observadome.

LaDon watched as both scientists looked to Barton with expressions of guilt and fear. Jendall seemed to have stopped breathing. Phelix answered for them both.

"Yes, Mr. Urthorn?" Phelix asked with

hesitancy.

"It's time we show them," Barton demanded.

Chapter 16

Have Man Will Travel

"Show us what, exactly?" LaDon looked toward Phelix with disappointment and a ton of curiosity.

"Well, you see, when we first discovered the idea of interspatial travel, we, um..." Phelix hesitated as he searched for the words.

Barton cut in, "LaDon, listen. Try not to be upset. You must understand—"

"I'm not upset. I'm just feeling a little in the dark while being held responsible for a project that has the ability to save the entire planet from its imminent doom. Now, I believe I am about to learn information that might very well change the course of my plans. So, let's not say I'm upset. Let's go with curious. How about that?" LaDon said with plenty of sarcasm.

"I'm going to explain. Will you relax and give me that much?" Barton's raised voice matched LaDon's sarcasm.

LaDon took a deep breath and attempted to calm his nerves. He noticed his knuckles turning white as he gripped the seat in front of him. He quickly realized his anger level. He mentally started with his fingertips, and then relaxed his hands, his shoulders, and so forth. His mind regained its calm, and he felt his respiration return to normal. Even though he was regaining his composure, he wanted an explanation of the truth quickly. He realized the magnitude of his outburst and attempted to reel himself in.

"I apologize, Mr. Urthorn. That outburst was unprofessional, but I hope you can understand," LaDon said with a calm but still somewhat heated tone.

"It was a well-warranted expression of your feelings. I might react the same way if I were blindsided. Hopefully, after I explain, it will be clear." Barton walked toward LaDon and continued with the explanation. "You see, when Jendall and Phelix originally came to the Assembly with their discovery, we immediately quarantined them both in order to decide which nation deserved the credit. We saw this was not an ordinary find and must be announced to the masses at a gathering. We also knew we needed to prep the Lead Representatives to ensure they had time to cross examine us. We had to take all proper precautions. During the meeting, someone mentioned the idea of sending a Solayan into another universe."

"Let me guess. Joh Lin?" LaDon interrupted.

"Who else?" Barton said with a smile as he continued. "This question caused us to open a side project. We wanted to see if this was possible. There was already proof that the Solaspheres could weather the journey, so we wanted to know if a Solayan could make the journey. When the question was raised again by our beloved Joh Lin at the gathering in the Prime Hall, we answered no. This was because no Solayan had yet to make the journey. Since then, actually directly after that meeting, we tested it. We placed them in a specially designed suit built to weather the confines of space. We sent a Solasphere in ahead of them to be sure no harm would come to them other than the elements. They were heavily examined by our medical teams both before and after the journey. We could find no difference in their vital signs except for a slight adrenaline rush which was explained away by one Solayan in particular. Phelix, how did you put it?"

"It was like being injected with a stimulant and seeing a loved one that has passed all at the same time. There was joy, elation, and fear all rolled into one," Phelix said with a slight grin.

Barton turned his attention back to LaDon and continued, "This discovery did not need to be brought to the surface just yet. We hoped we would find a world completely uninhabited by intelligent life. We would simply exist there. In light of recent events, I feel using this technology is necessary to complete our mission. Don't be upset with Phelix. He was under strict orders from me not to divulge this technology until it was absolutely necessary. This discovery was to be brought up at a later date, once this fiasco was behind us."

Aleen, who had been listening just a few steps behind Barton, spoke. "LaDon, dear, the last thing we want is to deceive anyone. That is never our intent. Some information is better left unsaid until necessary."

LaDon stood up straight to address Barton and Aleen as well as the other Assembly members in the room, "I completely understand. My outburst was premature. I should have trusted you. I should have trusted you all. Maybe it's the stress, maybe it's...I don't know. Either way, I understand now and hope you can forgive me."

"Like I said my friend, it was well warranted. I know how it feels to be in charge of something and feel as if your hands are tied behind your back." Barton stepped in closer and whispered loud enough for everyone to hear. "So, do you want to try it?"

"Me?" LaDon exclaimed.

"Of course! Who better to alter history than a historian?" Barton said with a smirk.

"You mean you want me to put on some type of suit and be whisked away by whatever magic is going on behind that bright light I see just before the Solaspheres disappear?", LaDon asked with an obvious amount of excitement in his voice.

"You'll love it," Phelix encouraged.

"It feels like you leave your body for a second," Jendall said from his terminal where he'd been hiding from the drama.

"You mean you've been too?" LaDon laughed in amazement as he looked toward Larissa. "Larissa, don't tell me..."

"Nope. This is all news to me, but I want to try it as soon as possible. This sounds exciting!" Larissa drew her fists into her chest and shivered with excitement.

"So, let me be sure I am clear. We are going to

dress up like these Earth people and insert ourselves into strategic roles within their nation. Then we are going to attempt to implement changes that will eventually mold their history to reflect our own. All in the hopes of guiding them to the technological discoveries we have uncovered. In the end, we will be able to tell them who we are without them thinking we're crazy?" LaDon said while looking back and forth between the faces in the room to see if what he was saying made sense.

"Exactly," Blaine said, smiling from a distance.

LaDon looked around Barton toward Blaine and saw his face beaming. LaDon could not help but smile as this idea. It sounded just as intriguing as it felt saying it out loud.

"And we're going to help you every step of the way, LaDon," Alex said as he joined LaDon on the main floor of the Observadome. "We just need to pick which points in their history should be adjusted. As we get closer and closer to our goal, we will find it easier and easier to remap the timeline beyond the point of the change. Since their technology will become more advanced more quickly, we can gather our information faster. Basically, they will be doing the work for us."

"What do you mean they will be doing the work for us?" LaDon asked.

"Look at it this way. Let's say we change something at a certain point in their timeline. After the change, if we enter their timeline at some point in the future, we can download all of their historical information that has occurred. This will catch us up and help us see the effects of our change. At each change, we will assess our situation and make decisions from there. We must be careful and make small adjustments. Larger adjustments will take time. This might mean one of us staying on the planet for years at a time, just to get ourselves in the right position to make a change." Alex finished giving time for someone else to speak.

"Isn't adjusting a timeline dangerous? How will we predict the outcome? Also, aren't we playing with their fate? Wait, let me guess. You've already thought all this through, right?"

Barton stepped in. "Dangerous? Yes. Why is it dangerous? Because we will not be able to make a one hundred percent prediction of the outcome. Their fate is death. Obliteration of their entire species. Why? Because they destroy themselves. Not because nature selected them to die. Not because the planet could not sustain them. To be blunt, it's because of their own stupidity. Because of one person with an itchy trigger finger, the entire world suffers. If you knew your entire species was going to be wiped out of existence, would you mind a little intervention?"

LaDon could hardly disagree with this logic. They were all going to die anyway. This would actually be saving their lives and prolonging their existence.

Barton continued, "The time frame after the World Ender is pointless to consider. We need this planet healthy."

"Joh Lin's not going to like this. All this talk of

changing their fate and inserting our own set of morals and technology? He's simply going to say no. He's not even going to entertain the idea." LaDon said after thinking it over.

"Let me handle Joh Lin. I've been debating with that Solayan for years. He will not be an issue. Our job is to protect this planet and its people. This will happen whether he whines about it or not. I know the other representatives will agree. Once again, Joh Lin will be outnumbered. He'll come around, though," Barton said with confidence.

"With all due respect, Mr. Urthorn, please do just that. And keep him out of our hair. Another Vaknoreeyan, me specifically, working so closely with a project that he will obviously not approve will definitely cause tension between the two of us," LaDon said in desperation.

"He won't be a problem, LaDon. You have our support in the matter," Blaine reassured him.

"So, shall we begin?" LaDon looked toward his team.

With their expressions on alert, the team looked to LaDon as they awaited orders. LaDon looked toward the Assembly members as he began to contemplate a plan of action. He watched as each of them made their way back to their seats. He began thinking through everything that had just transpired. *We are going to make ourselves look like these Earth people and infiltrate their social structure. Then we will make specific changes in an attempt to mold their history to reflect our own. While doing this, we will slowly guide them to the technological discoveries we have uncovered. Then, in the end, we will be able to tell them who we are without them thinking we're crazy. All of this while, at the same time, avoiding the demise of their species?*

After a long silence, LaDon said, "Well, in order to edit the history, we must know their history. Alex, you had the idea of downloading their history from their electronic archives. According to the information we have gathered, toward the end of their existence, they are somewhat technologically advanced. I believe this will be the first task for one of us to tackle. We need to go there and find a way to transmit their electronic data into our own database. This way we will be able to save time and Solasphere trips into their past. We already have tons of information, but I feel we can cross reference what we've learned thus far with a download of their history."

"So, who gets to go first?" Jendall raised his eyebrows with interest.

"That's simple. Me. Without question," LaDon said. "We will all get our chance, but I'm pulling rank on this one. I am becoming more and more attached to this idea of zooming across whatever void is out there and onto this planet. We will want to send a Solasphere ahead of me to assure no one will see me appear out of thin air."

"That sounds like a plan," Larissa answered. "LaDon, I have an idea. I feel we need to match our time variable with whatever unit of measure the people of Earth use to measure time. Let's find out what unit of time matches one orbit around their sun. Basically, the same way we measure a given year."

"Brilliant, Larissa! This might make it easier when we start to adjust their timeline," Jendall agreed.

LaDon immediately followed Jendall's words. "My thoughts exactly, Jendall. Larissa, I think that is a great idea."

"All right, we need to derive just how we are going to get this information back to Solaya. Do we port in and take one of their computing units? This way we can learn how they store information. Maybe even—" Phelix paced as he brainstormed aloud.

Alex cut him off. "No. We cannot remove any devices from the world. That in itself would register as a change. We must be careful of what we change. Even something as insignificant as a missing computing unit could lead to horrible ramifications. We *must* be mindful of this." Alex put a little extra oomph in his voice to drive his point into their brains.

"Then somehow we need to make a copy of the information. Maybe even a copy of the actual machine." Larissa said.

"I've got it." LaDon jumped to his feet. "We simply replicate it. Using a cloaked Solasphere, we can find a suitable location containing one of their computing units. Next, we record the area for the length of time needed to create a sufficient copy of the unit. This way we know there will not be any humans around to see the replication occur. Finally, we send one of us in to retrieve the replicated machine and immediately port back. If we coordinate this well enough, it should work. We're not removing anything and the Solaspheres will be invisible. The only time we would be visible is the time it takes to retrieve the machine and port back." LaDon realized he hashed out the entire plan without thinking about it. "Since that idea came out of nowhere, I need one of you to find an issue. What could go wrong?"

"I think it will work," Jendall said.

"Let's do it!" Phelix moved toward his terminal to prepare a Solasphere.

The team began by placing their Solaspheres in and out of buildings, structures, and anything they could find to find a suitable location. After about an hour of searching, Jendall popped his head up from behind his terminal.

"I've got it. I was sending units in and out of office buildings thinking that was the best place to look. I thought to myself that all we need is a replica of one of their units. Right now, all we want to know is how they store data. So, I decided to try their dwellings. I found one that is totally empty for an entire day. Since the replication process should only take a few hours, I figured this is the place," Jendall said excitedly.

"Nice find. Let's do it," LaDon said.

Jendall entered the necessary coordinates as Phelix brought over the Solasphere recently equipped with a replicator.

"The Solasphere will scan the device, which should take a matter of seconds. It will then turn to an open area in the room with sufficient enough space to create the copy," Phelix explained as he attached the device to Jendall's workstation.

The Solasphere glistened with its normal ray of blue light and vanished from sight. It appeared moments later as expected. The team turned to LaDon.

"Well, are you ready?" Larissa looked worried.

"Don't worry about him, Larissa. The journey is spectacular. No harm will come to him, I guarantee it," Jendall said as he stepped around from behind his terminal.

Jendall and Phelix led LaDon to the center of the room. "So, where does this happen, exactly?" he asked.

Suddenly, LaDon heard a loud crash from directly above him. A Solayan sized enclosure slowly lowered from the ceiling. It eased down and slowly surrounded LaDon.

"You mean, I go like this? I mean, dressed like this?" LaDon said to the thick container walls.

He recognized that Jendall and Phelix could hear him as they smiled and nodded.

LaDon looked through the glass out toward the Assembly. They were all standing in order to get a better view of the event. Blaine was standing, smiling from ear to ear, with his arms folded. Barton stood grinning proudly like a father watching his son conquer the latest sporting event. Aleen was leaning on Barton with her hands clasped in the crook of his arm. Alex was leaned forward with both hands on the table, watching intently, smiling like a sly dog.

LaDon's vision began to distort. The image of

those around him began to falter and then regain their clarity, just to be lost again. During the moments of visual clarity, he looked toward Larissa, whose face seemed to show a hint of worry. LaDon smiled and waved. She smiled back weakly.

Just then, LaDon felt as if he had been split in two but without pain. In what seemed like an instant, his world flashed blue, then black, then back to blue. After the final blue flash, the entire scene changed. He took a slow breath through his nose. The smell was unlike anything he had ever experienced. He tried to compare it to anything in his memory. Nothing came to his mind.

Immediately, after gaining his senses, his nose finally detected something familiar. The smell of a freshly replicated piece of hardware. He looked down toward the ground and saw a black, rectangular box sitting in the middle of the floor. He looked around the room. He noticed odd objects sitting atop the furniture in the room. The room seemed tidy except for the strange box sitting there, obviously out of place. *This is obviously what I am here to get,* LaDon thought as he reached down and retrieved the box. It was somewhat heavy, but nothing he couldn't handle. He adjusted the box under his arm. With his free hand, he retrieved the return module Phelix had given him just before he left.

He started to press the button to return home and thought back to something Jendall said about this place. *No one around for an entire day.* LaDon slowly removed his hand from his return module. Being mindful not to move or disturb anything around him, LaDon began looking at all of the interesting things in the room. He saw familiar things such as a bed, desk, and even pictures sitting on top of the desk. He leaned in closer to see the smiling faces in the pictures. *The pictures seem to be printed on some type of material. Interesting.* He continued his adventure around the room. He noticed objects organized neatly on a shelf. All similar in size, covered with the strange writings of the people of Earth. He took one of the objects from the shelf and began to examine it. He saw it was a paper-like material, all bound together. Between each piece of material, he saw more of the strange writing. He noticed some patterns and realized what he was seeing. It was information written in one of their languages. He was tempted to return with one of these items since it seemed to contain information. In fear of disturbing anything, LaDon thought twice and placed the item back on the shelf. He stepped back to the place he appeared and reached once again for his return module. With a slight press of the button, the world began to distort once more. It was as if he was feeling all the feelings he felt before but in reverse. After the flashes of blue and black, he started to see the familiar surroundings of the Observadome. Everyone was still standing in the same positions as when he left just moments before. The encasement slowly lowered around him and into the ground. In a few quick steps, Larissa was at his side.

"Are you all right? Did it hurt?" Larissa frantically ran her hands over his body to make certain he was still intact.

"I'm fine. I'm just fine. It was...exhilarating." LaDon's tone stayed calm, but a grin slowly made its way across his face.

Chapter 17

The Right Man for the Job

"So how was it?" Jendall asked as he retrieved the box from LaDon's grasp.

"Unlike anything I've ever imagined. It was as if I separated inside for a moment. My thoughts were one place and my body another." LaDon tried to explain but was still overwhelmed by the experience.

"Well, it looks like the plan was a success. Let's get this unit over to the lab and have our team examine it. The quicker we know what's inside this machine, the quicker we can understand how to obtain an electronic copy of their history." Alex took the machine from Jendall and hurried it out of the Observadome.

The team gathered around LaDon. He began sharing stories of his short visit on Earth. What seemed like a few minutes turned into an hour. He described the sounds, smells, and sights he experienced in the tiny room. He described the vibrant color of the walls. He talked about the soft flooring beneath his feet. What he discussed most were the unfamiliar smells in the air just before noticing the unique aroma of the replicated device. The team pounded LaDon with questions. They were obviously excited about experiencing their own trips.

"Why don't we take a midday break? Maybe by the time we finish eating, the team examining the unit might have some answers for us," LaDon said, trying to break the onslaught of questions.

"That sounds great. We can just continuing talking in the cafeteria." Jendall hopped to his feet.

"I think we've exhausted all the possible questions, don't you? I mean, I feel like I am starting to repeat myself." LaDon hoped Jendall would get the message.

"Oh, okay," Jendall said jovially. "I'll just make my own memories."

"That sounds like the best plan."

They all exited the Observadome laughing at Jendall's excitement.

"Would the Assembly care to join us?" LaDon asked as the team exited the Observadome.

"Thank you, LaDon, but we have someone bringing food to us. Certain Lead Representatives want an update. We all have a boss, I suppose." Barton answered for the Assembly but seemed distracted.

"Very well. We'll be back soon. Hopefully when we come back, the team will have something for us." LaDon tried to catch Barton's eye.

"Yes, let's hope so." Barton returned to his work without meeting LaDon's gaze.

Trying not to read too much into the exchange, LaDon pressed on to catch up with the group. They were holding the lift for him.

He's a busy man. Being distracted is normal.

Having to prepare for the onslaught of Joh Lin is even tougher, LaDon reassured himself as they entered the cafeteria. They each chose a meal and found an uncrowded area to sit. This was done strategically due to the sensitive nature of their discussions. Although it wouldn't have been a total loss if someone overheard, it was just better to keep certain things out of the public eye until there was something to actually report.

As conversation ensued, more talk of LaDon's tiny adventure continued through Jendall's persistence. Phelix and Larissa also had their share of comments as they relived the event moment by moment. Just as LaDon felt himself getting weary of the conversation, Larissa moved in a little closer to him and took his arm.

"I'm glad you're safe. I know they said over and over that nothing would happen to you, but a girl can't help but worry." She squeezed his arm tighter.

"As I said before, I'm perfectly fine, but it is nice to know someone was back here on Solaya who cared enough to worry." LaDon stared into her blue eyes, where he always found peace.

They leaned in just long enough for their lips to touch before an exclamation startled them apart.

"Ugh, it's not like he jumped in front of a moving transport or was attacked by a wild beast. He was in the hands of controlled, absolute science," Jendall said, pointing his nose in the air and snapping his finger toward the group.

"Leave them alone, Jendall. She was worried

about her sweet, adorable Solayan," Phelix cooed while laughing at Jendall's performance.

"You two are incorrigible." Larissa smiled as she tossed a bread roll at Jendall.

This caused much laughter at the table. LaDon appreciated the moment of peace amid chaos. The planet was doomed, and yet this group of individuals found room for laughter and even love. Unnoticeable to anyone but Larissa, LaDon squeezed his arm tight around Larissa's hand. It caught her attention as her face showed evidence of his affection.

He would have given just about anything to be alone with her at that very minute. Over the last few months, he realized just how close they had gotten. From being nervous outside his transport where they first shared a kiss, to spending time together, week after week, alone in his home. He realized this relationship was right where he wanted it to be. A beautiful Solayan who cared for him, surrounded by good friends. What more could he ask?

Just as his mind drifted to one of his favorite moments alone with Larissa, their lunch was interrupted by an oddly familiar face.

"Mr. Grafter?" The messenger spoke apologetically.

"Yes, I'm Mr. Grafter.", LaDon said as he sat up and Larissa dropped her grasp.

"I am very sorry to interrupt, but you and your team are being summoned back to the Observadome. The message I am supposed to deliver is 'the data is ready.'" Then the messenger scurried away into one of the corridors.

"Wow! How could they have interpreted the information so fast? I thought it would at least be a few more hours. And sending us a message in person? That could only mean they are in the middle of something. A meeting, perhaps?" Phelix exclaimed as they all began to rise from their seats.

They placed their disposables into the receptacle. LaDon checked the time and realized they had been in the cafeteria for a while. *At least we had time to eat,* he thought

They entered the huge double doors. LaDon immediately heard the clamoring of voices. None of them registered at first, but he could definitely tell the voices were new to the Observadome.

LaDon and his team made their way to the main floor in the center of the room. Extra seats were woven between each Assembly member, filled with the Lead Representatives. *And now Barton's worried face makes sense,* LaDon mused as he approached the Assembly's table.

"Mr. Urthorn? You sent for us?" LaDon asked respectfully while regally nodding to show recognition to the Lead Representatives.

"Yes, LaDon. I apologize if we interrupted your meal, but the representatives have some questions for you and your team. We have shared the latest events as well as our plan to edit the timeline. Also, we have decoded the information from the computing system you brought back. Its configuration was quite elementary. We have teams working on a device to interface with their technology. We can insert it into their computing systems and download information from these units. Also, it appears this species was able to develop a vast electronic network full of information. We believe this network will give us all the information we need to formulate a timeline of the planet's history. Between this information and the numerous recordings we took earlier, we should have enough information to derive what moments in their history we need to change in order to shape their future to match our own. That's where—" Barton was cut off by a voice to his left.

LaDon looked toward this voice and recognized the face immediately.

"That's where you come in, Mr. Grafter." Joh Lin's distinctive voice rang in LaDon's ears.

"Ah, Mr. Lin. So nice to see you, sir," LaDon greeted his representative respectfully.

From the side of his eye, LaDon noticed a displeased look creep onto Barton's face and slowly dissipate with apparent concentration.

"Yes, yes. Very good. A pleasure as always. To the point, from what I understand from the Assembly, you have already ventured to this planet, LaDon. Am I right?" Joh Lin lifted his eyes from the tablet in front of him.

"Yes, sir. That is correct."

"I am still not sure if I agree with this whole plan, but after hearing the story from Barton, I am beginning to ask myself, what choice have we? We need this planet. Would you agree, Mr. Grafter?" Joh Lin finished in what seemed like one continuous breath.

"Yes, sir. You are exactly right." LaDon answered as swift as possible again.

"What is your opinion here, Mr. Grafter? Do you agree that mucking around with this species' existence, just to satisfy our own, is a good idea? Not to mention sending our own kind through this time bending light show." Joh Lin looked up from his tablet again, awaiting an answer.

"To save this world, while at the same time saving another species from total annihilation, seems like a worthy cause to me, sir. To address the second part of your question, I trust the two scientists who developed this method of travel. They're brilliant. They're well trained, and they enjoy their craft," LaDon said with confidence. He felt Barton smile without having to look at him.

"Hmmph, a Delnokeeyan and a Kalloneeyan, huh? At least between you and Ms. Sonne, we're headed fifty percent in the right direction." Joh Lin obviously wanted a reaction from LaDon.

"I would die for this team, sir," LaDon said with a bit of disgust. "With respect, Mr. Lin, the wars are over."

Sitting back calmly in his seat, with a smile that hinted toward delight, Joh Lin slowed his speech, "Yes, they are. They most definitely are, Mr. Grafter. And who better to know this than the planet's most noted historian? You picked a good one, Barton. I'm impressed. He's got a backbone."

Barton directed his gaze up and down the table addressing the representatives, "Are there any other representatives that have questions for Mr.

Grafter or his team?"

All of the other representatives nodded firmly in affirmation of the plan put in front of them.

"You're Pomph's grandson, aren't you?" Joh Lin asked unexpectedly, scooting forward in his seat.

"Yes, I am." LaDon answered with a curious tone as his heart debated feelings of defensiveness.

"He was a great man. One of the most determined, and, may I say, entertaining Vaknoreeyans I have ever come across. The way he told his stories was almost like being there in person. We all felt it when he passed. I feel even more confident knowing you are a product of such an individual." Joh Lin finished once again leaned back in his seat.

"Thank you for those words, sir. I think about him every day," LaDon answered as growing serenity replaced his look of disdain.

"If there's nothing else, let's get to it." Barton looked to LaDon for the next move.

"Very well, Mr. Urthorn. Let's start with the information gathered from the team who examined the box we replicated. How long will it be before they can fashion this device that will allow us to download the needed information?" LaDon looked across each member of the Assembly.

Alex Cuberly answered, "They say they should have something replicated within the hour."

"Let's use this time to equip ourselves with Unifiers. I want these Unifiers fitted with every language on the planet. When we speak, the device should be set to relay the language used by the

Earthling we encounter. Also, I want to examine the development of this species a little closer. Let's see if we can get some insight into why they think the way they do. Furthermore, I want whatever data we have so far. Lay it out in chronological order. We need to find a time period in which to focus the changes. We need to pick a time where these changes make sense. We cannot hand a cave dweller a Solasphere and say, 'Here you go, we're from Solaya.' This means we need to look more toward a time period closer to more advanced eras. Phelix, if you will work on the Unifiers, Jendall, Larissa, and I will begin working on what period we feel meet this criteria the best."

The team scattered to get to work as LaDon finished.

LaDon could hear conversation between the Assembly and the Lead Representatives. Fortunately, the one conversation he was interested in the most, he was able to overhear.

"He's damn good. Let's just hope this planet's history is similar to our own. It will be easier to manipulate, don't you agree?" Joh Lin's distinct voice echoed behind him.

"Yes. That's a good point. We will know more when we can retrieve a clearer picture from the downloaded data. Where are those people anyway? They should be done by now." Barton responded, looking toward the entrance as LaDon looked away and stopped eavesdropping.

A few hours passed as the team finished up their research. They sent Solasphere after Solasphere, gathering as much data as possible, watching the Earthlings evolve, both physically and socially. Still waiting for the storage devices to be complete, LaDon noticed the Unifiers had arrived. Alex and Phelix proceeded over and took one device for each of the team members.

"Here you go." Alex passed out the Unifiers to the team. "It will pick up the language being spoken and translate it to our tongue. It will also translate your voice so they will be able to understand you. According to these specifications, they have designed it to interface with your viewers as well. This way you can choose which specific tongue you wish to be heard should you have the need. This is just in case you are in a room full of individuals who all speak different languages."

This was the first time LaDon had held a Unifier. It was smaller than he had imagined. He rolled the unit around in his hand. It was contoured to fit nicely in his ear. It's smooth, flesh colored surface felt a bit cold to the touch.

"Thank you, Alex. I'm sure these will work nicely." LaDon placed the device inside his ear. "We are finishing up here. Just a few more Solasphere recordings to put in place."

"What have you found out so far?" Alex asked.

"Well, it seems this species developed much as we did. Their earlier evolution consists of cave dwellers, archaic communication, and later evolving into upright, intelligent beings. We didn't spend much time in the earlier years. We have tightened our focus on the later years of evolution. It seems they become more intelligent around this time period." LaDon stretched out his finger to point toward one of the terminals. "Here."

"These structures can even been seen from high above the surface. One of the Solaspheres we placed in orbit around the planet gave us astounding results. This is how we gathered pictures of these surfaces from high above. Without going through every last item we've uncovered in the last couple of hours, we do believe we have found certain points of interest. We will know more when the team arrives with the storage devices. For now, here's our plan. We have mapped out the perfect locations to insert the device. We will place three of us at strategic points on the timeline and plug the devices into their systems. Since we do not know exactly how they will function, I cannot make an accurate plan from that point." LaDon looked past Alex toward the Assembly.

Just then, a team of unknown but familiar faces entered the Observadome. *This is obviously the news we are waiting on. Finally.* The unknown team approached the Assembly. Their leader was holding a simple square container. They sat it down in front of Barton. Barton opened the container, looked toward LaDon, and nodded. Barton continued his conversation with the unknown team. LaDon took his cue and moved toward the table. Barton pushed the box toward LaDon, who reached out and retrieved it.

Just as LaDon picked up the container, a voice from the team spoke, "Mr. Grafter, these units are designed with one purpose in mind. They will gather as much data as possible. They have been programmed to infiltrate anything stopping it from obtaining its goal. The Earthlings' computational skills are much like ours a few centuries ago. These devices will have no problem scouring their network for the information. They are also designed to pick up words from Earth vocabularies that represent historical importance such as news events and historical records. Simply connect these to a device that is connected to this network of knowledge, wait approximately five minutes, and then the device should turn green. This means its memory banks are full. Once it turns green, retrieve the device and return to Solaya. Each of these devices hold enough information to fill thousands of Solasphere memory banks."

"Thousands! Very well. Thank you. We're on it." LaDon said to the team member as they turned and left the Observadome.

LaDon looked back at his team. With a new brightness on their faces, he could tell they were eager to journey into this other world. At the same time, he could sense they were hiding their fatigue.

LaDon turned to the Assembly table and spoke, "Assembly and representatives, I realize we are on the cusp of even more great discoveries, but I feel we would all serve better with a little rest. We've been going at this hard, especially today."

"Good call, LaDon. I believe we could all use the rest." Joh Lin slowly stood from his seat, stretching at the waist, showing his age. "LaDon, I have to say, just the little bit I've seen today, this planet has hope after all. You show signs of a great leader. With a great team at your side, we are sure to survive. You're right. Let's all take a step back. Take a breath and return to our homes. We will start fresh in the morning. This all right with you, Mr. Urthorn?"

"By all means. LaDon, you and your team are doing a superb job. Let's reconvene tomorrow. We will get the information loaded into these devices and into our main database and go from there," Barton replied as he stood from his seat as well.

"Thank you. We will begin bright and early tomorrow." LaDon stepped over to his team.

"Thank you, LaDon. I was getting toward the end of my rope as well. Although, I would be able to find the energy somewhere if it meant visiting this planet. I can't wait!" Jendall said as he and the team gathered their things.

"See you in the morning, LaDon." Phelix said as he drifted toward the exit. "Thanks for calling it when you did. I think we're all appreciative."

As the room began to empty, Larissa approached LaDon with her sweet, adorable smile.

"Would you care for some company tonight?" Larissa asked seductively. "I promise not to keep you awake too long."

"I believe I would enjoy the company very much, but only if you stay the night." LaDon smiled slyly as they exited the Observadome.

Chapter 18

The Law is Absolute... and so is Murphy

After a relaxing night alone with Larissa, LaDon was anxious to execute the new plan. He was very happy that Larissa spent the night. He was going to need his strength for his next venture. He knew it would require much thought and careful planning. Their relationship couldn't have come at a better time in his life. In both of their lives probably, if you were to ask Larissa.

LaDon and Larissa arrived at Nalkalin. As he exited his transport, he waited next to it for her to join him before they went inside. They entered the facility together just as they had done many times before. Everyone was used to seeing them together. Their relationship was no secret. On separate occasions, LaDon and Larissa each received a brief lecture by Aleen Fabian. LaDon didn't see it as a lecture, but more as advice from someone that was going through the same thing. The entire planet knew that Barton Urthorn and Aleen Fabian were inseparable. A workplace relationship could prove to be iffy, especially if one party was the supervisor and the other the subordinate.

"You must be able to make the big decisions

without your feelings for Larissa clouding your judgment," Aleen had said.

In any case, Aleen was very specific with her wording. She expressed to both of them that, in the end, she wanted them to be happy and that she cared for them very much.

As they entered the Observadome, LaDon saw the normal hustle and bustle of people. This was a different sight from his first few weeks at Nalkalin. When just Phelix, Jendall, and Larissa occupied the Observadome, things were much simpler.

LaDon and Larissa parted ways and she wandered toward her terminal. LaDon paused because he noticed something going on overhead. He looked up to see someone high above working on the third of three glass enclosures.

"Ah, they have installed new transport modules. Looks like we're just about ready," LaDon said to anyone listening as he placed his things beside his terminal.

He looked toward the Assembly table. Each member's head was buried in their terminals working diligently on their own tasks. Barton looked up from his work, eyes unfocused until he met LaDon's eyes. With his normal swagger, Barton gave LaDon a look of approval and slowly looked away. LaDon felt confident in today's plan as he eased his way over to Jendall's terminal where Phelix and Jendall were talking.

"Are we ready?" LaDon asked. The pair of scientists seemed extra eager.

"Ready as we'll ever be, boss," Jendall said

enthusiastically. "I can't wait to try this!"

"You're getting too worked up. It's not that great," Phelix said, trying to burst Jendall's bubble, which was all but impossible.

"It was quite spectacular if you ask me." LaDon smiled toward Jendall to encourage his excitement. "On a serious note, have we fully scrutinized the three locations worthy of our task?"

"Yes. It didn't take us long to find three suitable locations. One is an office building at night, one is a vacant home, and one is a building referred to as a *library*." Jendall answered promptly. "We sent in reconnaissance Solaspheres to each location ahead of time for eight minutes. That should be long enough to get in, get the data, and get out. Everything should go smoothly."

"A li-br-ar-ee.", LaDon sounded out the word like a toddler learning its native tongue.

"Yes. It translates to data-housing. The old kind when everything was stored on parchment," Phelix explained.

"I know what data-housing is. It doesn't take a scientist to know that," LaDon joked. "Very well. Nice job, my friends. I will make a small announcement to the Assembly and we'll get started."

LaDon approached the table and began to speak, "Pardon me, members of the Assembly? Do you have a moment? Before we send the team off for data gathering, I would like to fill you in on some details."

"By all means, LaDon. Please, go ahead. We are listening." Blaine reacted like he was programmed to respond when the Assembly was addressed in such an official manner.

"We have performed reconnaissance on three specific locations to deploy the team. Each member is instructed to arrive on the scene, insert their device, wait until their unit's light turns green, and then return safely to Solaya. If there are no questions, we will begin the procedure." LaDon paused to give the members time to respond.

LaDon watched as Blaine looked down the table at each Assembly member.

Once each Assembly member gave the go ahead, Blaine looked toward LaDon and said, "the Assembly has nothing to add. Please proceed."

LaDon turned toward the team. They were already preparing themselves for the journey. He made sure each member had a return module and their storage device. LaDon reviewed the plan once more to make sure everyone was on the same page.

"All right. Let's go over this one more time, shall we? We don't want any mistakes," LaDon emphasized as the enclosures lowered to the ground floor. "You've each watched the Solasphere recording sent in ahead of you. You know where the unit is located. Port in, insert your device, wait for the light to turn green, grab the device, hit your return module and come back home."

The team nodded as they entered their enclosures. LaDon walked over to his terminal to activate the sequence. With a small whir and a flash of light, each member disappeared almost simultaneously. In less than a minute, the same sound began again. With a flash, each enclosure was filled, except the enclosure which contained Larissa. LaDon paused for a moment as he stared at the empty enclosure.

His mind attempted to avoid the obvious by imagining the whirring sound was still going and Larissa would appear at any moment. As the enclosures lifted from around Jendall and Phelix, they both emerged, excited and ready to share their stories. Jendall immediately rushed to Phelix with jubilation. Both scientists were unaware of Larissa's empty enclosure. Just as LaDon felt himself losing control, he sensed the Assembly coming down from around their table.

"WHERE IS SHE?" LaDon yelled. "Where the HELL is she, Phelix!?"

Phelix slowly turned from an excited Jendall only to see the empty enclosure.

"NO!" Phelix ran to LaDon's terminal to see the last commands entered. "The readings here are correct. She went to the designated coordinates. It's all right here, LaDon."

With looks of disbelief, Phelix and Jendall started to approach as a withering LaDon fell to his knees beside the enclosure. He was slowly coming apart as Barton appeared by his side.

"Just step back!" LaDon bellowed as he collapsed further into himself, curving his back forward and shrinking further into the floor. "Just step back!"

"LaDon, listen. Listen to me." Barton rushed in close to LaDon, placing his arm around him to shelter him from the onslaught of pain. "Take a deep breath. You've got to think clearly."

Through staggered breaths, LaDon responded, "What's to think about? She... didn't... come... back. The rules say you must return to the same point in time in which you left right after you transport. So either she lives there until she dies or she dies immediately. Either way, she's dead, Barton. How do you expect me to think clearly?"

Barton lowered his head and sighed. Just as he was about to speak, LaDon lurched from his mound of depression and grabbed the closet Solasphere.

"LaDon, don't. You don't have to..."

"Yes—I have to. I have to see what happened. I'll just send in this Solasphere to the point when she arrived on Earth. I *have* to know," LaDon said, hoarse with rage.

LaDon connected the Solasphere. Through blurred vision, he input the coordinates and sent the cloaked Solasphere to investigate. The Solasphere disappeared and returned as expected. LaDon grabbed the device and furiously connected it to his terminal.

"LaDon, you don't have to watch this. There's no telling what you'll see. Just..." Phelix's pleading reached deaf ears.

LaDon sat up in his chair, still sobbing, looking for any type of release. The Solasphere recording began. At the beginning of the recording, the Solasphere turned to face Larissa since it was programmed to do that when noticing points of interest. LaDon got a small sigh of relief when he saw her. The Solasphere moved into position to get a layout of the room. The room was totally empty. With the environment recorded by the Solasphere overtaking his senses, mixed with his shuddering insides, he felt uneasy as he watched her attach the storage device. As the light turned green, he watched as Larissa remove the device and place it in her pocket. She took a step back and started to reach for her return module. Just then, he noticed her eyes change as if she noticed something off into the distance. She took a few steps toward a desk and leaned over to view a picture frame. *What are you doing Larissa? Just leave!* LaDon mentally screamed at her as he watched her continue her curious adventure through the room.

Suddenly LaDon heard a familiar rush of sound and a click. It was the reconnaissance Solasphere they had sent in earlier disappearing from its mission. LaDon looked back toward Larissa and saw that she also was startled by the noise. She took a few steps away from the items on the shelves and reached for her return module once again.

Just then a noise registered on LaDon's Solasphere. Naturally, the Solasphere moved to investigate. As it moved closer to the noise, LaDon shifted the view point of the recording over in Larissa's direction. She had heard the sound as well. *Hit your return module! What are you doing? Get out of there!* LaDon thought as he watched helplessly. As the Solasphere made its way toward the noise, it passed through the doorway and pointed itself down a short corridor. LaDon saw a shadowy figure moving toward the room. The Solasphere's view of Larissa was directly adjacent to the view of the shadowed figure. He watched as Larissa moved toward the doorway and the shadowy figure moved toward the room Larissa was occupying. The figure obviously knew there was something in the room. LaDon's heart was racing as his mind formulated the rest of the recording, but his desperation to know the truth kept him from stopping the recording early. As the Earthling reached the doorway, LaDon heard Larissa scream as two loud pops registered in the Solasphere's audio sensors. He watched as Larissa's body fell loosely to the floor like a rag doll. The recording went blank.

LaDon came out of the recording gasping as if he had been holding his breath the entire time. He crumbled into a thousand pieces, crying in agony, slipping from his seat and onto the floor once again.

"This can't be! They said the place was empty. They said the place was empty!" LaDon repeated as he lost all sense of control.

Through his disarray, LaDon could feel hands grabbing him and pulling him back into his chair. He could hear voices talking to him, but the only voice making it through the haze was Barton.

"LaDon, listen. I need you to listen to me. Can you do that for me?" Barton demanded. "LaDon, can you hear me?"

LaDon nodded slowly, attempting to regain his composure. With each calming wave, he fought off another outburst.

"Good. Hear me when I say this." Barton leaned down eye level with LaDon. "We can get her back."

With his head still swimming, LaDon looked up to find the entire room encircling his chair.

"What? How do you mean get her back?" LaDon said without the ability for logical thinking. "She's gone. We're not that far advanced. We can't stop death!"

"Maybe not in this timeline," Barton said with a calm, all-knowing inflection.

LaDon looked up to see Barton smiling sheepishly. Jendall and Phelix were wearing the same expression. LaDon felt the reasoning center of his brain coming into focus, trying to make sense of Barton's comment.

"Not in this timeline?" LaDon looked up confused, still feeling the fatigue of his sorrow.

"It's tricky but it might work." Jendall jumped into the chair at his terminal. "All we need to do is send someone back at a different location in the home, stop the attacker just before...and get both parties out of there, fast!"

"What's the tricky part?" LaDon asked from his chair surrounded by the Assembly and Jendall.

Jendall explained, "Here's the trick. LaDon, this is more than just a trick. This is a scientific law. Listen very carefully. If Larissa tries to return to the exact time and place she left, in our timeline, the calculation says she will be erased because we went back for her. Think with me here. To us, here at this point in our time, she's gone; therefore, we must save her. If we save her and she returns to us five minutes ago, then we never had a reason to go back to save her in the first place. Time will not allow this paradox to occur. She will simply vanish. However, when we go to save her, we must give her a new return module, specifically designed for her. She must use the new module to return to the exact moment here on Solaya that we go back to save her. Does this make sense?"

"Yes. The same reason you can't travel through time in our own universe. It's contiguous. Since their timeline is separate from ours, being its own universe, we can edit from a distance, so to speak," Phelix explained.

"You are exactly right, Phelix. Do you follow what I am saying, LaDon?" Jendall looked to LaDon, moving his head to meet eyes with him.

With the Assembly still gathered around him for support, LaDon proved he was now thinking logically again. Barton and the rest of the Assembly took a few steps back to let the team work.

"I'm going." LaDon said immediately.

"Wait, can't we just send a Solasphere? Can't it just give her new instructions and a new return module?" Jendall protested.

LaDon's head was firmly wrapped around this idea. He wasn't taking no for an answer.

"No. It *must* be one of us," LaDon insisted, "There is just not enough time for her to realize what's going on, take the new instructions, understand them, and use the new module. She's in a tight spot, Jendall. No one should be expected to figure all of this out so quickly. One of us must go in, foil the attacker without harming them, explain to her what's happening, give her the new module, and return to Solaya."

Phelix rushed to his terminal and began fiddling with the return modules.

"LaDon, give me just a moment to configure her return module. It shouldn't take but just a second...there. Done. Remember, this module is specifically for her. You will have your own. *Do not forget.* She cannot use the one she currently has." Phelix explained as he handed both devices to LaDon.

"All right, LaDon. I am placing you right behind the attacker, just before it reach the doorway. You'll have just a few seconds when you arrive. Get this right the first time. It will get really confusing if we have to send you back more than once. You would end up watching yourself die or watching yourself screw up. It would get very disorienting, if you could imagine," Jendall explained from his terminal as he input the coordinates for the center enclosure.

Barton stepped in for a final word. "LaDon, whether you hear me right now or not, remember, the most important thing. Do not harm the attacker. Disarm them, knock them out, or whatever it takes to stop the attack. Do not, I repeat, do not kill them. We are not here to disrupt the timeline any more than we must. We're here to save, not destroy."

LaDon heard Barton's words as he nodded in agreement and stepped into the enclosure.

Barton took a step back. Aleen grasped Barton's arm as LaDon heard the familiar whir of the teleport. His heart began to race as his vision distorted. After the familiar rush of adrenaline, he realized he was right behind the attacker. He was amazed just how close he was to the shadowy figure. Luckily, teleporting from Solaya to Earth wasn't preceded by flashes of light and whirring sounds. With a quick swing of his fist, LaDon caught the shadowy figure just above the ear. The attacker's head jarred to the side and bounced off the wall. The figure dropped to the floor.

LaDon quickly stooped down to examine the Earthling. He assumed the being must breathe, so he checked for any signs of life. The figure moved slightly, moaning incoherently.

By now Larissa was approaching the commotion, trying to make sense of the noises coming from the dark. LaDon saw Larissa, elated that she was alive, and reached for her. He was immediately stunned by a fist to the face. He quickly realized Larissa was swinging furiously, trying to defend herself. LaDon grabbed her arms, screaming her name frantically. She slowed her attack and paused to look.

"Larissa! Larissa! Stop!" LaDon said as he held her arms tightly. "It's me. LaDon!"

"LaDon! Why? I mean, what? What's happening? Why are you here?" Larissa's twenty questions came like a flurry.

Blindly overcome with emotion, LaDon grabbed Larissa and pulled her tightly to his chest.

The smell of her hair and the warmth of her face began rushing through his senses. He felt himself about to break into a tearful frenzy and forcibly held his composure.

"What's wrong with you?" Larissa said as she backed away from LaDon's grasp.

"I love you, Larissa Sonne! I love you more than I've ever loved anyone in my life. Do you hear me?" LaDon looked into her eyes, now more visible since his eyes had adjusted to the dark surroundings.

"I love you too, LaDon, but you need to start talking fast. I am more than confused." Larissa smiled, half love and half confused out of her mind.

"There's no time to explain. We need to get back to Solaya." LaDon glanced back at the Earthling who was regaining consciousness.

"But this human here," Larissa said with lingering confusion in her voice.

"It's fine. Just knocked out," LaDon explained, hurrying her back into the room.

"Okay. Please explain this to me when we get back," Larissa said as they prepared to port back to Solaya.

LaDon reached into his pocket to retrieve Larissa's new return module. What seemed to occur in slow motion, from his peripheral vision, LaDon saw Larissa retrieving her original return module.

"NO!", LaDon screamed which startled Larissa into a small, fury-like state.

"All right, now it's just annoying. What the hell, LaDon? What now?" Larissa asked as LaDon snatched her original return module from her hand.

"Use this one. This module was designed specifically for you. It will return you to the moment I came to get you. You can't go back to the original time you left," LaDon explained.

LaDon watched her face change from annoyed to aghast. He could read the look of horror forming on Larissa's face.

"I never came back, did I? That's why you're here. That means that I...that Earthling must have... and you came to save me." Larissa's eyes filled with tears as her intellect slowly revealed the truth to herself. "You didn't see it, did you? Please tell me you didn't watch me die?"

LaDon's eyes began to well up with fresh tears as a new wave of emotion overtook him. Larissa could see the pain on his face. He fought back the emotion, trying to remain sane for the both of them. Just from the expression on LaDon's face, she was able to extrapolate the truth. She threw her arms around him, sobbing into his neck as the reality of the situation was thrust upon her. He gently pulled her away so he could see her face. He wiped away her tears as she attempted to put on a brave smile in the midst of chaos.

A small rustle came from the corridor just outside the room. He looked over to see the Earthling slowly regaining consciousness and trying to push up from the ground. LaDon quickly tucked her old return module in his pocket, separated himself from her embrace, looked to her, and said, "I love you, Larissa. Just press the button. Let's go home."

Chapter 19

There's Work to be Done

As Earth faded from view, the familiar aroma of the Observadome started to fill his senses. LaDon scanned the neighboring enclosure for Larissa and felt a sense of relief when he saw her. She was safe and sound back on Solaya where she belonged. Phelix and Jendall both rushed to Larissa's side to greet her. LaDon noticed the tears in her eyes had not subsided as he had expected. She smiled at Phelix and Jendall for a moment, but then they were met with rage from a very distraught Larissa.

"How could you let him watch? There's only one logical explanation for what just transpired. He came back and I didn't. So, someone had to find out what happened. The only way to do that is by Solasphere, right? Am I right?" Larissa scolded Jendall and Phelix as they lowered their heads in submission.

By now the Assembly had made their way down to the floor only to be witness to Larissa's fury. Barton tried to step in and calm the situation with his wisdom.

"Larissa, it's hard to explain..." Barton attempted to defuse the situation but was drowned out by Larissa's continuing onslaught.

"He had to take a Solasphere and connect it. Next, he had to send it back and attach to the recording. Finally, he had to sit and watch the scene play out. You mean to tell me you couldn't stop him? Maybe even watch it yourself?" Larissa attempted to contain her fury, which caused her to wither into tears.

As everyone in the room tried to think of a way to console Larissa, all eyes turned to LaDon as he rushed toward her side. He knew he was the only one that could bring her down from the agony.

"Larissa? Larissa, look at me," LaDon said firmly as he pulled her up to his eye level. "An army couldn't have kept me from sending that Solasphere. I was in a total rage. As a matter of fact, it probably rivaled how you are feeling this very second. It's not their fault."

"Oh, I know it's not their fault, LaDon." Larissa buried her face in his chest.

She cried what appeared to be one last outburst as her emotions started to reel themselves in. LaDon wrapped his arms around her as she took a deep, shuddering breath. She looked over toward Jendall and Phelix, who both watched with understanding looks. Larissa rushed over to them and wrapped her arms around both of them.

"I know it's not your fault. Not one bit. I'm so sorry. You didn't deserve that in the slightest." Larissa said as she looked at each of them through watery eyes.

"We are going through this together, Larissa.

Your reaction was justified. I would never blame you for having those feelings." Phelix's well placed words seemed to calm her a bit more.

"Besides, we are friends, not just coworkers. Friends understand each other. We spend so much time together. I consider us family." Jendall's heartfelt reaction shocked the room for a moment until the next words spilled from his mouth. "It's not my fault you died. If you hadn't been so curious and simply hit your return module when you were supposed to, none of this would be happening."

This comment resulted in a much-needed laugh from Larissa. She smiled at Jendall and turned back to face LaDon. She looked at him for a moment and then reached into her pocket as she walked over to Alex who was standing just behind Barton.

"Here you are, Alex. My storage unit as ordered." Larissa presented him with her unit.

"Very good. I will send these off with the other ones. We will compile this data and see if we can't come up with a timeline of events. From there, we will pick highlights from the timeline, send in some Solaspheres to record the events, and input the recordings. It should line up nicely and start to look like our own archives." Alex took the devices and walked out of the Observadome.

Blaine spoke in his familiar tone, "I believe at this point, we should take a break. Let's give the other team a couple of hours to compile the data. Let's reconvene in exactly two hours."

As each person went their separate ways,

Barton approached Larissa and LaDon. He walked up to them and placed one hand on each of their shoulders.

"I'm glad you are both safe. You gave us quite a scare there, Miss Sonne. This goes without saying, but Jendall was right—as strange as that may seem," Barton said with a smirk. "Curiosity won that battle. Let this be a lesson to us all. Now, I must ask you. You had to attack the human, yes?"

LaDon answered quickly, "Yes sir. I knocked the human unconscious. I checked its respiration, assuming they breathe as we do. When I did this, I realized the human was still moving. This gave me confidence that it was still alive. Finally, just as we ported back, I noticed the human climbing to their feet. From this, I believe they survived. I would like to think they would have suspected us as being nothing but mere thieves or intruders."

Barton looked on with an expression of concern, but then relaxed his face. "Well, once we compile this first round of initial data, we will know a lot more about this species. In the meantime, the other Assembly members and I have been discussing when, where, and even how we plan to edit this timeline. Have you given this any thought, by chance?"

LaDon looked at Barton with disbelief and said, "It's so strange that you ask this question. Actually, I have, but I am still kicking around an idea. I will know more when we study the timeline in more depth over the next few hours."

"Very well, then. I believe we'll hold our ideas

until then as well. What we find out just might change our current mindsets altogether." Barton started to dismiss himself from their company.

"Sir?" Larissa said as she caught Barton's arm before he got away. "I want to apologize for my outburst earlier. It was disrespectful to you, the Assembly, and everyone else in the room."

"Miss Sonne. Your friends said it best. It was a moment of passion. No one will be faulted for such actions," Barton said.

LaDon felt it couldn't have been said better.

"Would you care to join us for something to eat?" Larissa said to Barton, which piqued LaDon's interest.

"Well, um..." Barton stopped to think.

LaDon whirled around to offer his invitation as well, "We would love for you to join us, sir. Actually, I would love to hear anything you might know about my grandfather that I may not know."

Barton's face went soft as he glanced toward LaDon. "You know, I think that sounds like a wonderful idea. Although why don't you two join me?"

"In the Assembly's kitchen? What a treat!" Larissa said with a chirp in her voice.

For the next hour and a half, LaDon and Larissa's ears were filled with stories from the past. Stories of political struggle, adversities, and even unknown friction between certain representatives which had been resolved long ago. For Larissa, hearing talks of the past was fascinating as she got a glimpse into real-life politics and hardcore debates. A much-needed escape from the mundane affairs of science.

For LaDon, the experience was completely different. Hearing stories from the past, especially stories involving Pomph working in close vicinity with Barton, made LaDon feel somehow even more connected to Barton. Seeing another side of the man he'd admired for so many years. It was like meeting a long distance friend that you've known for years but were just now seeing them in person for the first time. All the while, he listened to the stories, filling in the missing pages as if he was putting together a puzzle before knowing the picture.

"He was and will remain a great Solayan, LaDon. Please, don't ever forget that," Barton said as he finished his plate and sat back in his chair.

"I can guarantee that, Mr. Urthorn. He lives with me everywhere I go. He's the voice in my head."

"He would be proud, LaDon. Very proud." Barton met LaDon's eyes. "So, shall we head back? Hopefully they have something for us by now."

Just then Barton's consciousness drifted into his viewer. For a few seconds, LaDon and Larissa awaited his words. The latest information, straight from the man himself.

Barton's eyes came back to reality as he looked at both of them with a smile and said, "Perfect timing. They're ready for us."

The three stood up from the table and made their way back to the Observadome. As they entered, LaDon could see the team responsible for compiling the data. A few of them were from his old team in the historical center.

"LaDon!" A voice from the distance crossed the Observadome.

"Ramil!" LaDon bellowed as they hurried to meet each other.

The two met with an embrace as they shared the normal pleasantries.

"So this is where they shipped you off to. Everything happened so fast!" Ramil exclaimed.

"Hah! Yes it did. So how are things? Anything exciting happening in the world of Solaya?" LaDon asked casually as Larissa walked up beside him.

"I wouldn't know. I've been working on compiling the historical data you guys have obtained over the last six months or so. I knew you were involved, I just didn't know what section." Rami's eyes were distracted for a moment. "Oh my. And may I ask your name, Miss?"

"The name's Larissa. Larissa Sonne," Larissa said courteously. "LaDon, I'm going to get with Jendall and Phelix and see if they have any ideas on our next steps."

As Larissa walked away, Ramil watched as her hand separated from LaDon's. Not to mention the longing look LaDon gave her as she walked away.

"It doesn't take a mind reader to get that one. Nice work, LaDon. You've got it all over here. Solaspheres, other planets, humans, working directly with the Assembly, and a gorgeous lady friend." Ramil seemed to enjoy his role as Captain Obvious.

"Yes. I would say I'm quite happy. Glad to see you haven't lost your talent for bluntness, Ramil,"

LaDon said with a knowing smirk.

"Wouldn't trade it for anything! On a serious note, I want to tell you something before I go." Ramil leaned in to whisper to LaDon, "After breaking down this timeline, I've noticed something. This planet's inhabitants are eerily similar to us. Same body type, facial features, and they even breathe the same air. Also, their societal structure is almost identical to ours. Well, to a point. We were talking about this. It's really strange, LaDon. They even have separate factions. The humans call these factions cultures. I realize this is the first planet we've ever come across containing intelligent life, but the odds are ridiculous. Without seeing other planets and lifeforms to put this data up against, one would have to assume our physiological makeup is the ideal balance for intelligent life or something like that. Straight out of science fiction, I tell you!"

"Thanks for the tip," LaDon said as his conspiracy theorist friend wandered off into the crowd.

Although the things Ramil told him were quite helpful, LaDon didn't have time for theories and whimsical thoughts. It was time to get down to business.

LaDon made his way to the Assembly table, which was now crowded with his team, the historical research team, and of course, the Assembly.

"Everyone, please, may I have your attention?" Blaine's voice rose up above the clamor. "We have the timeline mapped the best we know how. It is about to be displayed across the dome. The beginning of this timeline starts at the beginning of intelligent life. It ends at the point when we gathered the data. Please look up at the displays. I have a layout of the timeline."

The Observadome's magnificent viewing capabilities went to work immediately. Only taking up half the Observadome, Blaine configured the timeline representation to stretch across the viewing area.

"Now listen closely, Solayans. This is the entire plan from beginning to end. I want to rehash it one more time, so pay close attention. We will record the specific events marked on the timeline and derive an appropriate edit to parallel our own history. Along the way, we will place specific technological advancements. This will aid in stopping them from destroying themselves as well as the planet. Teach them Jendall and Phelix's new science and reveal the truth. The truth of who we are, why we are there, and what we have done. Is this clear to everyone? Please, speak up now; otherwise, we assume you understand the critical nature of this mission." Blaine finished as he looked around the room for questions or suggestions. "Very well then. It's settled. If the historical compilation team will exit the room, our other team will get started with the Solasphere recordings of the specific moments in time."

Everyone knowing their roles, the historical team exited the Observadome. Ramil gave LaDon one last goodbye as he exited. LaDon smiled in Ramil's direction. LaDon then turned to his team, who all seemed anxious to get started, and laid out his simple plan.

"All right, I see we have thirty specific points. Each of us will take seven. Whoever finishes first will take the remaining two. Agreed?"

The team agreed as they divided up the responsibilities and began sending Solaspheres They each made quick work of their seven recordings, gathering the data needed to devise an edit. Jendall finished up the last two as the team gathered around his terminal to watch. Jendall punched in the last few numbers to store the recording appropriately. The team turned to LaDon, who then turned to the Assembly for the next move.

"Assembly members? We have gathered the specified information. I would like to request time to study this material in depth. I feel with a full understanding of their history, we can make a better informed decision on when and where to attempt these changes," LaDon said with authority.

"How long do you feel you will need, LaDon?" Barton asked without hesitation.

LaDon took a moment to think through his answer, *We are talking thousands of years of information. Although well compiled and laid out, not only is this thousands of years, but thousands of years of mental evolution. Evolution of the nation, evolution of their ways of life, and all of the changes that go with it. Well, all we have is time, I suppose.*

"I would ask we take at least two months, sir," LaDon answered firmly.

"I agree. Two months minimum. We have a lot of ground to cover. We will continue to monitor progress from the Observadome, but I want these next two months spent in study. We've got to get to know these humans. What drives them? What do they want? Do they learn well from their own history?" Barton continued but one thing distracted LaDon.

With a raised brow, LaDon nodded in agreement and looked to Larissa while still speaking loud enough for all to hear, "Do they learn well from their own history? That's the key. Getting them to realize that war is not the answer is critical to this entire endeavor. It's what made Solaya thrive. Once the nations of Solaya realized that war would simply destroy them all, things began to change. Solayans began to change. Underground peace treaties were formed simply from the fear of self-annihilation. Scientific breakthroughs were made simply by joining forces with one another. But from that came the wars of competition and pride. Maybe this race of humans are better than that. Maybe they could teach Joh Lin a thing or two."

LaDon finished his thought and turned back toward Barton. LaDon could tell Barton had been smiling during his words to Larissa.

"So, are we clear?" Barton looked toward LaDon, whose mind slowly rejoined the conversation.

"Yes, sir. Perfectly clear." LaDon realized he had been caught drifting in his own thoughts.

"Be glad I was rambling, Mr. Grafter. You didn't miss anything important," Barton said with a smile as LaDon realized the rest of the room noticed his little trip to the other side of his mind.

"My apologies, sir. It's just something you said about history repeating itself. It is a well-known fact, but one I feel is overlooked more often than we realize. Solaya learned during the Forgotten Wars. They learned that fighting amongst themselves was killing them and depleting their natural resources. And for what? Pride? They learned that fighting wasn't the answer. They were rewarded for this behavior as their scientific shortcomings were fulfilled by their unions with their enemies. When they came together, the reward of scientific improvement made the peace easier, and even stronger in some cases. If we can replicate this in Earth's history, I feel we will succeed. Giving them the technology is not the answer. Let's make them earn the technology. That's the key. If they earn it, they'll respect it." LaDon stopped as he realized his thought made even more sense when spoken aloud.

"Mr. Grafter, that brilliance is exactly why you are heading up this project!" Barton exclaimed as he looked to the other team members. "LaDon said it, my friends. We must mold this world. Teach it. Help them understand that what Phelix saw in that Solasphere recording doesn't have to happen. War is not the answer. Life is too short for such nonsense. We can all be different, yet strive toward the same goal. There is a Solayan from each nation represented in this room, and I have grown attached to each of you in one way or another. We need each other. We've learned this. Now we much teach them. Let's give them life, while sustaining our own."

Barton's powerful words of inspiration made

their way into the hearts of everyone in the room. As he delivered his words, each Assembly member had taken their place beside Barton to support him. As LaDon's team took in Barton's speech, they each returned to their terminals to start their journey through time. LaDon couldn't help but feel a sense of purpose. A sense of duty. A sense of heroism. It was his job to save them. It was up to his team to save these humans from themselves. Not to mention, to save their own world from its inevitability.

Chapter 20

The Potter and the Clay

Months went by in deep study. Each team member began breaking down the timeline from their own unique perspective. Phelix and Jendall focused on the scientific advancements of the culture. They began sharing with the team certain similarities between this species and their own.

"It's so interesting. It seems their technological leaps forward are the direct result of the people's needs. In some cases, their desires." Phelix shared as he entered some notes into his terminal.

"You know, I noticed the same thing. It seems when they want to make things easier, someone finds a way to make it possible. Sometimes they even improve on it," Jendall said as he leaned back in his chair to ponder the concept. "Much like here on Solaya. I also notice that this planet's technology doesn't develop any further than our own technology level before the Forgotten Wars."

LaDon was taking a completely different approach. He saw Earth's history through the eyes of a historian, watching cultural exchanges, religious debates, and even wars involving both culture and religion. He related to Phelix and Jendall's findings, but from a completely different perspective altogether.

"I tend to agree. I believe this will be very

valuable information. That's exactly the part of the human mind we need to focus on once we start introducing technologies. It seems a nation tends to refresh its mindset after a great struggle. I suppose after dealing with a struggle, the sigh of relief might cause this type of effect, not only for the individual, but for the nation as well," LaDon agreed as his thought was cut short.

"It seems their advancements to science are the direct result of their willingness to progress and change. That is the main reason I do what I do. Everything can be explained if given enough time," Phelix stated.

"It seems we are both saying the same thing, or at least our thoughts have the same goal. As I appreciate your scientific viewpoint, my friends, naturally I have been looking at things through the viewpoint of historical progression. I began by watching them once they learned how to shape metal. I figured that would be a distinguishing point since they are starting to see more scientific concepts and not so many religious debates. Although they still have their share of those from what I see. In any case, convenience is starting to play a factor by then, which shows social evolution as well. Watching humans of power rise and fall. Watching complete cultures extinguished by greedy dictators." LaDon sat up in his chair and propped his elbows on his knees. "There is something quite intriguing, though. When we put the information we've gathered up against the initial data we recorded through our Solaspheres, they don't seem to line up just right. It seems they record their history through the eyes of the victor. One instance in particular. I studied about a people overrun by technology, while still very primitive, and forced to leave their land. Quite honestly, it was despicable to see. I'd love to hand those poor souls a few technological advancements of their own and watch those iron clad humans squirm."

LaDon finished his thought with a huff under his breath. To LaDon's surprise, Barton who had obviously been listening the entire time, approached the group with his own interpretation.

"I feel the focus will need to be on their ability to adapt to change. Our ultimate goal is to introduce to these Earthlings the truth about life on other planets. In our case, other universes. Do they take change well, throughout their history I mean? How do they accept it as a people? From what I have been studying, when a change occurs and it promises an ease to their way of life, they are more accepting of the change. We can use this to our advantage. As we work to introduce technologies as a means to end the war, and these technologies stem from a collaboration with their enemies, we should begin to notice changes in their history much like our own. Each side should see there is no benefit in fighting, wasting resources on global war, when joining forces would make them stronger. We have our own history to go by, and I intend to use this resource to its fullest potential," Barton explained with the normal enthusiasm.

Larissa began speaking at the end of Barton's

thought. "I agree, Barton. If there is one thing I have noticed, the minds of these Earthlings continuously change as they become more and more aware of science. Science is something that is constantly changing. It says here that they used to believe their world was flat, and if you sailed far enough, you would fall off the edge. Here it says a man was sentenced to house arrest for his entire life simply for stating that their planet revolved around the sun rather than the sun around the planet. It's almost funny when you say it out loud. It does read in both cases that religion played a huge part here. I guess scientific proof isn't always the answer to a problem. At least not right away."

The team agreed with low murmurs and nodding heads as they pondered the information each of them had contributed. LaDon felt positive about the fact that his team had so many different viewpoints as they reviewed this planet's history.

A few more days passed until finally a full two months came to an end. The team's heads were now filled with ideas, but most importantly they now contained the knowledge of this new planet and its people. They started to reference points on the timeline with actual dates used by the Earthlings.

"They seem to use a lot of different calendar types to explain dates and times. This one here, the Gregorian calendar, is more widely used toward the end of their existence. It seems an important religious figure set this into motion around 1582 AD," LaDon explained as he reviewed a particular article he fancied. "We will use this to measure time.

I have overlain these numbers to the timeline above us on the displays. If you look to the displays, you will see I have plotted important milestone dates along the timeline. The dates are now listed to correlate with these specific points in time. We have started the timeline approximately one hundred years before this cross over period. It seems as time progresses, their history records the date as getting smaller. Finally, when it hits zero, the number counts upwards. Without going into it, they refer to this break in years as BC versus AD. No reason to explain further. Let's just go with it. If you want an explanation, look it up."

"What's that very first milestone they marked? The only one before the AD mark?" Phelix asked, leaning forward. "It would almost seem senseless to attempt to change something so early in their technological adolescence."

"I agree. Remember that these were just suggestions. They did not have the time or information to study as we do, but we will humor these points nonetheless as we attempt our own edits." LaDon told Phelix as he examined the notes given by the other team concerning this point in time. "It says here that a great library was burned to the ground. The article refers to it as the Library of Alexandria. The articles that correspond to this event do not reveal who burned the structure, but it does say it contained information that could have advanced technology must faster than it did. Wait! Hold on. How would the other team have this information when all they had to reference were

THE POTTER AND THE CLAY

Solaspheres? They wouldn't have had the storage devices."

"Oh, you mean *that* first dot on the timeline there?" Jendall asked with a fake innocence.

"Jendall?" Larissa cut her eyes in his direction.

Jendall could not hide his laughter as the team looked at him with a mixture of frustration and humor.

"Well, it looked interesting, so I tried to sneak it onto the timeline and into their notes," Jendall said sheepishly. "It is very interesting to say the least, isn't it?"

"Actually it is. I tell you what, Jendall. For your amusement, let's find this library, stop it from burning down, and see what we get." LaDon looked toward Barton for approval. "If something goes wrong, we will port in just after our change and simply burn it down ourselves. That will really make your article ring true. They will have no idea who burned it down."

LaDon looked back to the Assembly in time to see Barton laughing in approval.

LaDon continued, "Send in a Solasphere to find it. Find how it burns and see if you can stop it."

"Already ahead of you. I found it a few days ago. I sent in some cloaked Solaspheres to gather the information. I know just how to stop it. Don't worry, don't worry, I didn't change anything. It was my own private project." Jendall pulled up the information in his system.

"Hmmph, I really don't approve of sending

anything over without approval, Jendall," LaDon scolded.

"All right, all right, I understand. I won't do it again, I promise. You still want to try it?" Jendall asked with no loss of enthusiasm.

"Of course we can, although I want to go on record to say that this will be introducing a great change into this timeline. I feel generally the changes should be more subtle, smaller, and less intrusive," LaDon said with a smile. "Either way, we can always undo it. Make the change, and we will reevaluate the timeline around the same time we gathered the information using the storage devices. If anything goes wrong, send a Solasphere in right after you stop it from being burned and burn it down anyway."

Jendall fired up his terminal and began his work. Obviously he had this all thought out as LaDon watched him work. Jendall shot over a Solasphere and stood from his terminal.

"Be right back." Jendall grinned as he marched to the enclosure lowering itself from the ceiling.

"Be careful, Jendall. Do you need one of us to go with you?" Phelix asked anxiously.

"Not at all. Once I figured out who burned it initially, it was quite humorous. It was a child who was sneaking around at night, carrying a dimly lit piece of wax. Who knew, right?" Jendall snagged his Unifier from his terminal. "Oops, can't forget this."

From inside the enclosure, Jendall nodded to Phelix, who executed the commands Jendall sent him just before approaching the enclosure. Within seconds, Jendall disappeared. Seconds passed and he reappeared. To the surprise of everyone in the room, Jendall returned laughing wildly. He gasped for breath. The team gathered around him to ensure he would survive the onslaught of laughter.

"What? What could possibly be this funny?" Larissa said, trying to contain her own laughter in response to Jendall's hysterics.

"I told him the creatures of the night were coming for him, so he better run home!" Jendall stammered through the sentence trying to regain his composure. "I have never seen someone run so hard in my life. He even dropped the candle. I simply stepped on the flame to extinguish it and ported back."

Phelix and Larissa listened as Jendall replayed the story one more time of the poor boy running for his life. LaDon immediately moved to his terminal and started connecting a Solasphere. Alex and Barton appeared behind LaDon, startling him a bit, as they inquired about his actions.

"What do we have here?" Alex asked with a quick, low tone.

"Oh! You startled me.", LaDon said in surprise. "I am getting this Solasphere ready to see the effects of the change. Why? Is there a problem?"

"No. No problem at all. It's just nice to see your mindset. Being more interested in the actual effect of the change than performing an edit," Barton explained with a proud tone in his voice.

LaDon smiled as he executed the command. The whir of the transport broke up the commotion between Phelix, Jendall, and Larissa. They all turned to see Barton, Alex, and LaDon all gathered around LaDon's terminal. At the same time, they saw an empty Solasphere enclosure flash, obviously porting over a cloaked Solasphere. This brought them back to reality a bit as the Solasphere returned as quick as it had left. LaDon felt around for the Solasphere. Once he recognized the familiar shape, he connected the Solasphere to the terminal, threw the video to one of the monitors above, and started the video. The video started out dark at first, slowly adjusting to the surrounding light.

"There's not much light in this area. Where did you put it, LaDon?" Barton asked quietly as he watched the view screen.

"It's supposed to be outside on a sunny day in the middle of a busy area bustling with people," LaDon explained as he looked at his previous notes.

As his eyes wandered back to his terminal to view the notes alongside the video playing, he noticed something in the sensory data.

"The air quality. It's almost unbreathable," LaDon exclaimed in disbelief.

"Where are the people, LaDon?" Larissa asked.

"How should I know? I can tell you they can't breathe, though. That's for certain," LaDon said without a shred of evidence to prove any of the theories racing through his mind.

"Wait, let me see the air quality readings." Phelix rushed to LaDon's terminal. "Just as I thought. It's the same."

"The same as what, Phelix?" LaDon leaned

back in his chair to let Phelix have a closer look at the terminal.

"When we first visited Earth after the nuclear war. Remember, the black curos in the air. That's the same readings I got the first time."

"But look at the date, Phelix. That can't be true." But LaDon couldn't deny what was right in front of him. "But that would mean..."

"Exactly. Not burning down this library caused a chain of events which increased their technology so rapidly, that they still destroy themselves. Only this time it happens three hundred years sooner? Maybe even earlier than that?" LaDon grabbed a tussle of his hair. "Well, this is a simple fix. Jendall, give me the exact times and coordinates you used."

"Why, LaDon?" Jendall asked, not quite paying attention. He obviously still wondered what just happened.

"I am going to go in right when you left and burn the place down," LaDon explained with an accomplished grin on his face.

Jendall punched a few buttons on his terminal and LaDon received the information he needed. He walked to the enclosure and nodded at Alex, who was still close to his terminal.

As Alex pressed the execute command, LaDon smiled and said, "I won't be long, I promise."

He disappeared and reappeared as designed. He said nothing as he moved toward his terminal. He connected the same Solasphere and sent it off again. The Solasphere returned and LaDon connected it to the terminal.

"I've sent this Solasphere to the same place as before. Let's see how everything looks now." LaDon looked up as the recording began playing on the screen above.

The Observadome filled with sounds of street cars, honking horns, and the shouts of humans as they moved about the busy city. LaDon checked the readings to assure himself everything was back to normal.

"There we go. All better. I believe we need to be much more strategic in our approach next time. Wouldn't you agree?" LaDon said with an exhausted grin on his face.

Jendall looked over to LaDon slowly and said, "I couldn't agree with you more. I believe I will leave history molding to you from now on. After all, you are the historian."

"Let's start on these other points on the timeline and see the effects of those smaller changes," LaDon said as he noticed the team was still pondering what just occurred. "I realize the outcome of such a change was drastic. This was actually the exact opposite of what we'd hoped. I want to look at these other points and see if we have better luck. Each time we go in, let's make sure we have a plan to undo the change. We've only got one planet to work with here."

The team began diligently working. Plan after plan, scenario after scenario, digging for some type of change that might cascade them into a greater plan. One that might lead them to a point where they could begin molding Earth after Solaya's history. Days and weeks went by as they attempted multiple edits trying to find something to start them toward their goal. Each point on the timeline eventually turned up as a dead end. the Assembly became heavily involved. Debates broke out as they hashed out ideas and tried new things. Quickly exhausting the timeline points given by the previous compilation team, LaDon's team created their own points. None of the edits yielded acceptable conclusions. For LaDon this proved almost maddening. *It seems no matter what we try, some adverse effect keeps the change from taking hold or simply not panning out like we plan.*

Each time the team attempted an edit, it failed, so they had to go back and unravel the change. LaDon began to continuously play a conversation over in his head. It was his conversation with Ramil, his old historian partner that was part of the compilation team. *They are extremely similar to us. Their physical makeup and their separation of nations.* While their physical makeup tended not to play much part in his thought process, LaDon began to attempt to expound on their short conversation. *They are very similar to us. They have Americans. We have Kalloneeyans. They have Asians. We have the Nuwee. At the same time, our goal is to structure their development to mimic ours. Just before they destroy themselves, their technology level is comparable to Solaya's just before the Forgotten Wars began.*

"Wait a minute!" LaDon's exclamation startled

everyone in the room.

"Yes, Mr. Grafter?" Blaine called from across the Observadome.

"Oh, I'm sorry. I didn't realize I said that out loud." LaDon apologized but continued quickly as to not lose his thought. "Nevertheless, I believe I just had an idea."

"What's that exactly, LaDon?" Alex asked.

LaDon realized he had a captive audience.

"I realize we're all frustrated right now. Try to hear me out here." He took a deep breath as if he was going to say everything in one rush. "Our goal here is to edit their history. We want to edit their history in such a manner that their technology advances enough to understand the science we are using. This way we can tell them who we are without being resisted. Right?"

The team nodded in unison as LaDon thought about his next point before speaking.

LaDon continued, "Well, we all agree the best way to do this is to mold their history to reflect our own. Phelix, at what point on the timeline is the point where the World Enders are launched?"

"According to our records, it states a little after what the humans would consider their twenty-second century," Phelix answered, still waiting for LaDon's reasoning to click.

"Has anyone been reviewing the current events of the day right before the onslaught or have we all been researching the past?" LaDon paused for a response but none was given. "Exactly! They're already at war! Phelix, can you tell me who launches at whom?"

Phelix hurried to his terminal. LaDon could tell his idea was starting to seep into the minds of his team as each of them returned to their terminals frantically to look up whatever questions were on their minds.

"Let's see. My Solasphere recorded the bomb being launched from this location around 2120 A.D. That cross references to this location. This area of the world was known as America for a long time. Around that time, this area was just recently taken by another faction from an Asian culture when the bomb was launched." Phelix's eyes grew worried as he recalled the devastation caused by the initial launch. "I'm telling you LaDon, it's all-out chaos by this time period."

"Like it or not, Phelix, this is exactly where we need to be," LaDon exclaimed. "Quick, someone look this up. Actually, everyone start looking. Who is fighting who? Who are enemies? Who are allies? Fine lines, gray lines, imaginary lines, I don't care. Are there any truces? Any cease fires?"

The team worked vigorously for the next few hours, compiling data and deciphering friends, allies, and foes. LaDon took the information given to him by each team member and began incorporating it into the new timeline he had lain out.

"Okay. Stop for a moment." LaDon displayed his latest timeline high above for all to see. "Look. It's right there."

"What LaDon? The timeline up there? It just looks like another timeline with more events,"

Jendall said at first glance.

"But LaDon..." Phelix said as he stood from his terminal as if to get a better view of the displays above. "The points on your timeline haven't occurred yet."

Phelix and the rest of the team stared at the timeline trying to make sense of what they saw.

"Watch," LaDon waved his hand in a flourish as if he was a magician performing his act.

LaDon placed a separate timeline just below the one he just created. They lined up identical to one another. He labeled his secret timeline as 'Earth' and the separate timeline as 'the Forgotten Wars.' He watched the faces of each member of the team as his idea landed on their faces. Their expressions changed with each passing thought. From curiosity, to confusion, to rationalization, to finally logic, their faces told the story in their minds even better than the great Pomph Grafter. As LaDon watched the understanding creep into their minds, he threw another volley their way. He placed a chart in the shape of a T next to the timelines. Side by side the words read, Americans - Kalloneeyans, Asians - Nuwee, Europeans - Vaknoreeyan, Africans - Nuweeyan, and finally Australian - Delnokeeyan.

"Of course! Right before the bombing, their nations are at war. Unbelievable! So if we stop the bombing..." Phelix paused, trying to think through the logic.

"History lesson. At the beginning of the Forgotten Wars, who fired first?" LaDon asked as he looked around the room.

"The Delnokeeyans, of course," Aleen said. LaDon smiled at her. "It was recorded as one of the most devastating blasts in history. Definitely an event we Delnokeeyans are not proud to take credit for."

"You are exactly right, Ms. Fabian. And what wonderful advancement was the result of this attack?" LaDon looked around once again.

Barton stepped up to take a swing. "The Vaknoreeyan and the Nuweeyan had an alliance at the time. One of the very first alliances recorded, if I recall my history lessons correctly. The Nuweeyan people took the brunt of the hit. They knew they could not take another attack of that magnitude. It was recorded that the Vaknoreeyan and the Nuweeyan had just started sharing their sciences when it was discovered that you could create Collectors."

"Oh, you mean the deep space exploration devices, right? The device that absorbs thrust and other kinetic energy and redirects the energy for propulsion?" Jendall piped in, testing his historical knowledge.

"Yes. Precisely. Although space exploration was not the original intent. Am I getting this right, Mr. Grafter?" Barton paused and looked toward LaDon.

"Quite accurate, sir. Please continue."

"You see, the Collectors of that day were placed in strategic locations around the land. The next Delnokeeyan launch was absorbed and the shock was redirected into the air. So basically, nothing happened. So, the Nuweeyan people and the Vaknoreeyan people agreed to share this science with the other factions. Bombs became useless. It's even recorded that they shared the science with the Delnokeeyans. They were very impressed. They've always been proud, and they respect an admirable opponent." Barton finished his history lesson.

"So we allow the bombing to occur, but we go in prior to the launch and get two of their factions talking just enough to introduce Collectors, allowing them to stop the nuclear bombs from destroying the planet?" Larissa asked, following the logic.

LaDon answered almost immediately, "Exactly. Our history has shown this to work brilliantly. This recurring phrase continues its streak. History repeats itself. So, let's prove it true once again, shall we?"

Chapter 21

We're Going In

"LaDon, what you are saying is impossible. You can't just appear at some point in history, walk up to these humans, and say, 'Hey, how are you? What do you think about this convenient piece of technology?'" Jendall said in a somewhat humorous tone. "You would run into the same problem we face now. If we told them who we are and where we come from, they would think us mad. We would never get them to trust us."

LaDon smiled for a moment and looked toward Barton, who shared the smile. For the first time in his life, LaDon knew exactly what Barton Urthorn was thinking. Spending this past year working side by side, LaDon had benefited from Barton's wisdom. Maybe it was those long talks after everyone had left the Observadome. One conversation in particular came to LaDon's mind as he made his way to Jendall's side and turned to face the rest of the team.

"I had a long conversation with a close, personal friend of mine one night about this very issue." LaDon looked toward Barton once more as he continued his thought.

Aleen stepped up beside Barton, her face beaming with pride, proud of the man she loved more than anything in existence. From LaDon's perspective, it was obvious Aleen and Barton had discussed this very thing as well.

LaDon continued, "The Forgotten Wars did not end overnight. In fact, they lasted just over three hundred years. History records that alliances were formed only by certain factions within the nations. Not all Vaknoreeyan and Nuweeyan people liked the fact that an alliance formed due to the creation of the Collectors, but they could not deny the peace it brought. These antagonists were suppressed by those who believed in the changes. Either way, those who disagreed with the alliances could not deny the advantage of the advancements of their own technology. Also, they could not deny the benefits they saw from other alliances that began forming. They were getting stronger, and at the same time there was light at the end of the tunnel. They saw hope, freedom from war, and even a future where their children could live. These alliances were formed by none other than the representatives of the time. They may have served under different titles, but their positions stood the test of time, and later became known as the Lead Representatives. These representatives changed once or twice over the three hundred year period, but eventually the wonderful Assembly was crafted from these revolutionaries, and other representatives stepped in to take their place. It was the perfect balance. From there, our nations thrived. There are still a few stragglers here and there, digging up the past, trying to stir emotions, but there are too many level headed individuals in the world now. Such bigotry is simply a thing of the past."

Larissa cut in with a question. "But LaDon, this took hundreds of years to accomplish. Plus, each nation trusted their representatives with everything they had. Even with their own lives at one point or another."

"Yes, this is true. How did they gain their trust?" LaDon asked as he looked around the room.

"Time." Barton's one word caused a moment of silence.

LaDon took a shallow breath and leaned against Jendall's terminal, allowing the minds of his team to spin with frenzy. He watched as each of them processed just how they might accomplish such a thing with the knowledge at hand.

Phelix was the first to speak, "So this means we have to talk to them. Interact with them. Get one of them to make the right moves. But how do we do this without telling them who we are? It just doesn't make sense, LaDon."

LaDon stood from his casual lean and looked Phelix in the eyes. "We don't tell them."

"But if we don't tell them, then how do we change their minds? What are we going to do? Become their friends? Take them to dinner?"

Phelix stopped. His words began filling in the missing pieces for the other team members as well. LaDon waited with anticipation as he watched his team arrive at their own conclusion.

"You want us to be the representatives! You want us to become one of them, even live as one of them," Larissa interjected as the realization landed on her face. "It's the only way I can imagine this to work. We each pick a culture and introduce the two parts to every science as needed. If history does actually repeat itself, this should mold the course of their history, their economy, even their entire government to match our own. We make them like each other. This will work without a doubt."

Larissa walked off into the distance, muttering to herself, as LaDon took the floor.

"This is not going to be an easy task. Infiltration will be the most difficult part, but with a few impressive technological advancements, we should gain favor quickly, rise to the top, and pull these races together." LaDon said as he watched his team's faces begin to understand in unison.

"This won't happen in the blink of an eye. We are going to need to live there for a while." Jendall said as he squinted and smiled as if hiding a more interesting thought.

"Yes, at different points along the timeline as well. It will take careful planning, but I know we can do it. The invention of the Collector is a perfect place to start. It will save the planet from ruin and save the humans from destruction. From that point, their history isn't written. We get to write it and edit out what we don't like. This should also appease Joh Lin." LaDon shot a looks toward the Assembly who could already read his next comment on his face. "We can change whatever we like from that point. It's no longer their timeline. It's ours. They were about to die anyway."

LaDon looked over at his team, huddled

together already discussing this new plan. LaDon already had this plan well thought out a while back after his heart to heart discussion with Barton. It took Barton leading him to the inevitable conclusion that there was no other way to tell the people of Earth who they were without interacting with them in some form. LaDon had wanted to try every possible avenue, gathering as much information as possible, in hopes there was another way, but after today's series of events, he knew Barton was right. LaDon had even pushed this thought to the back of his mind until Ramil's words about similarities in relation to Solayans. This was the mission. This was the fate he must face. They must live amongst these Earthlings. Act like them. *Be* them. All the while, gaining their trust through technology and attempting to teach them that the power they lacked was in the hands of their enemy. There was no other way to obtain the political power they needed without killing, and that was simply out of the question. It was going to be an uphill climb, but LaDon knew if he could recreate the events just as they were recorded, the plan might succeed. All of the talk surrounding this plan stirred a few memories for LaDon. All of the years of listening to Pomph's stories came rushing back, fresh as ever. In a strange way, LaDon thought how he would actually be reenacting the very stories Pomph told him time and time again. His heart reached back for those memories with bittersweet happiness. Those memories would now play a key role in the coming years of his life. There was no happier place inside his head than when he was with his grandfather.

"Do we start today?" LaDon asked his team as they broke from their chattering huddle.

"Why not?" Jendall said as the team gathered around LaDon's terminal.

"I believe we all realize at this point the plan is going to stretch for years. We have to become human. We are literally devoting our lives to this project." LaDon explained as he guided their minds through the process he had been imagining over the last year. "Luckily we are similar in our physical makeup. Just some slight prosthetics should do the trick. Remember, if you ever find yourself in trouble, simply get somewhere safe and activate your return module. It will bring you back so you can adjust your issue and port right back at the point you left. Let's begin. I want to start five years before the first launch of the nuclear warheads. We stop that from occurring and return here to reevaluate the timeline and go from there. Agreed?"

The team agreed as they settled in for the journey of planning. They scanned the timeline, reviewing each culture's strengths and weaknesses. From this they each chose a culture based on their specific knowledge, skill set, and overall ability to perform the tasks at hand.

Time marched on as days turned into weeks, weeks turned into months, and months turned into years. The first edit took five Earth years exactly. Periodically they coordinated small trips back to Solaya to meet up and discuss their plans and any issues hindering them from weaving themselves into a nation. LaDon and Larissa made quite a few extra trips back to Solaya for other, personal reasons. At first Jendall and Phelix made their share of visits back to Solaya to visit family and friends. Since this project had slowly become common knowledge throughout Solaya, family members started to comment on how each time they came back for a visit, they could tell that they'd aged. Jendall and Phelix requested to start taking their families with them when they foresaw staying on Earth for extended periods of time. This didn't bother LaDon and Larissa since they ported back frequently enough to spend time together on Solaya. Since they could port back to Earth at the moment they left, the luxury of time was just that—a luxury.

Their lives became intertwined with the people of Earth. The culture, the history, and even the primitive views of science all started to make sense as they learned more about the lives of Earthlings. They found themselves even liking the humans, understanding their differences, and even caring for them at times when it seemed they were faced with staggering odds. With the first five years complete, they had lived amongst the humans, rising to a state of respect by giving a few hints to scientific circles. Teaching them small nuances of science, while at the same time, reviewing the timeline every now and then to see if anything major had changed. Small changes in science were minuscule enough that the launch still occurred on schedule. The team met back one last time before stepping forward to begin the first of many alliances, which would in turn unveil the invention of the Collector and give the Americans enough time to defend themselves, while making sure the entire planet saw it.

"Funny. I picked this culture for a reason. I wanted to meet the poor soul that launches the unscheduled nuclear attack. He's actually not that bad of an individual. He has a family and kids just like I do. Our girls have even played together. But he's scared. He has lost all hope in humanity. So many others have lost hope as well. The one person I was so distraught over years ago, knowing he still presses that button, I now feel for him. In fact, I almost agree with him. Such a strange toil of emotions, I just can't get that out of my head." Phelix lowered his head as Larissa consoled him.

"Well, friend. Here's the tricky part. You've got to let him press that button. He is going to be the first step in a long line of events that will shape his future and his children's future, forever," LaDon encouraged. "All right, let's pull our families out of there. We don't want to leave them alone over there if we don't have to."

The team made the proper arrangements, going back for their families and bringing them back to Solaya with new return modules. With their families safely home, the team began working on the next few edits of the timeline.

Larissa and LaDon were the first to attempt the formation of an alliance. The Europeans and the Americans, who had once considered themselves allies, hadn't quite seen eye to eye for many years since the late twenty-first century. Larissa and

LaDon had become well respected in their fields of science as well as for their political prowess. They exited Solaya and headed back to Earth to attend a scheduled event where the two cultures would be communicating openly with one another for the first time. LaDon and Larissa disguised the meeting as purely scientific in nature, stating they believed by comparing notes from their last encounter, which almost destroyed a small country, they just might have had something that could change history. With much disdain and scrutiny, they were allowed to meet, but only in a heavily guarded, populated area. This was just what LaDon had hoped.

LaDon stood from his seat, glaring at Larissa with the best scowl he could muster, and began the meeting by addressing his European colleagues present at the occasion, "I realize this is one man's opinion, my friends, but please, hear me out. These wars are costly. We are depleting our own resources as we tear at each other's throats. I believe both sides can agree to that statement. Now I know you wouldn't all be here if you didn't yearn for peace somewhere in your hearts. I appreciate everyone allowing this gathering and I hope we can keep it peaceful. Ms. Sonne and I are on a purely scientific mission. It took key people in both nations to pull together a meeting of this magnitude. I have shared my research with Ms. Sonne and she has shared hers with me. We believe we have formulated a way to absorb the shock of an incoming missile attack, no matter how sudden. We believe we can even redirect this shock into useful, kinetic energy to be used in other ways."

Gasps came from the crowd as murmurs began from both sides. Both LaDon and Larissa stood to calm the masses as each side listened to their own representative.

"And now, we are going to prove it," LaDon said quickly.

A small roar stirred once again as LaDon stepped toward the table.

LaDon raised his hand and out of pure curiosity, the noise lowered to a small rumble, "I ask for the American soldiers present here today to keep their weapons on me as I reach into this bag. Please everyone, do not be alarmed."

LaDon reached into his bag and slowly pulled out a grenade. Several clicks of guns from both sides of the room began echoing through the meeting room.

"Please, my good people. If any solider takes a shot, you will have to answer to authorities. Please, I implore you, keep your fingers off the trigger. This is not a war. This is a demonstration," Larissa said desperately. The peace in her voice soothed the anxiety of the soldiers as they relaxed their posture still keeping their guns at eye level.

"This is a grenade, as you can all see. Down there at the other end of the room, please notice the cone shaped object built into the wall. We call it a Collector. This is a smaller version of the real thing. I will pull this pin and toss this grenade toward the Collector. Also, please note the furniture to each side of the object." LaDon pulled the pin and tossed the grenade.

The soldiers trained their guns on LaDon as the grenade left his hand. The grenade landed far away from the spectators, and the soldiers lowered their weapons so they could focus on the event. Just as their weapons lowered to their side, the grenade exploded. The whir of the Collector filled the room with sound as the force hit the sensors. To the crowd's amazement, they felt no concussion from the blast, and the furniture didn't move.

"That furniture should be in pieces," a random voice came from one of the European soldiers. "There should be nothing left. Nothing!"

His words were heard throughout the room. Everyone was in disbelief, trying to make sense of what they just saw.

"I would like to ask one of the American soldiers to take one of their grenades from their belts and toss it toward the Collector," LaDon announced where everyone could hear.

The same soldier that spoke momento before stepped forward and pulled a grenade from his belt. He lofted the grenade as it landed almost in the exact same spot as LaDon's. The grenade exploded violently, the whir commenced, and the blast was absorbed without hesitation.

"Impossible." A reverent voice said from the lull of the crowd.

"No, my friends. Not impossible. I have been working on this technology for quite some time now. It wasn't until I noticed a trait in your defensive systems that I noticed the key to my mystery. With

Ms. Sonne's work combined with mine, we were able to make this possible. She has a brilliant mind. She is a credit to your culture and to your way of life." LaDon slowed his voice for a dramatic effect. "Americans and Europeans alike, please hear my words. We have been fighting for so long. We have been advancing our technologies trying to outdo each other. Just imagine what we can do if we put our heads together. This is just one facet of science. There are numerous fields each of us have advanced, trying to one-up the competition. I'm not asking you to end the war right here and now, but I am asking you to think about it. Think of the possibilities. Let me make one final announcement and we will end our demonstration. With larger versions of these Collectors strategically placed through both of our lands, we can stop the blasts of the largest bombs known here on Earth. Also note, the force of the blast will direct any harmful substances, even radiation, in the same direction as the blast. The Collectors are also charged to attract radiation and store it safely. We ask you keep this news inside our own cultures. Only Americans and Europeans shall know of this technology until it is unveiled by our enemies should they decide to attack us. May we attempt to use this technology to bring peace and not create more destruction."

The meeting ended as the crowd dissipated, leaving alone LaDon, Larissa, and their full complement of soldiers behind.

"You know this was televised, right?" Larissa gathered her things.

"No, I did not. How many people do you think that reached?" LaDon asked.

"Anyone who could watch restricted channels, I imagine. Luckily, according to Jendall and Phelix, it would have only been received by Americans and Europeans," Larissa answered with a pleased look on her face.

"That's perfect! Do you think it made a difference?" LaDon asked as Larissa stepped in closer to him.

"Let's go find out," Larissa whispered.

LaDon and Larissa both returned to their homes on Earth to be sure everything was clear. They made sure no one was around, and they each ported back to Solaya at the designated time. They were greeted by smiling faces of both Jendall and Phelix as well as the Assembly. The room was even filled with a few Lead Representatives who were all smiles as well.

LaDon stepped down from the pad as the enclosure rose into the ceiling. He walked up to Larissa, who was just stepping from her pad.

LaDon leaned into Larissa, and with a raise of his eyebrows, he said, "Let's see if it worked."

Chapter 22

Line 'Em Up, Knock 'Em Down

Barton greeted LaDon and Larissa as they returned from their first attempt at bringing two opposing cultures together.

"We are awaiting the arrival of the reconnaissance Solasphere to see if you were successful," Barton began as the room turned its attention to him.

"Here it comes now, sir," Alex called from one of the terminals.

Alex watched for a brief moment as the Solasphere played back the recording. LaDon and Larissa both looked at each other as if they were sure their mission was a success.

"Well done, everyone. Very well done, indeed," Alex said as he disconnected himself from the session.

"This is the first step toward a greater goal. Even though, for us here on Solaya, it might not seem that long, I realize the past five years of your lives have been spent with the people of Earth. By the looks of things, five years hasn't done much damage to your youth. Nevertheless, my friends, we are glad to have you back here for the time being.

Have you given much thought to our next phase in the plan, LaDon?"

LaDon answered immediately, "Yes sir. As you mentioned, when you've got five years to plan, you have time to give things a lot of thought. It's all we have been thinking about, honestly. These inventions have to come slowly. With all my heart I believe these inventions should come cloaked in the disguise of a reward. Almost like a reward for good behavior. It's exactly how it was viewed during the Forgotten Wars. Both sides would see the benefits of an alliance. It seems these Earth people have had to fight for a lot in their past. Maybe we give a new piece of technology each time they attempt talks of peace. We reinforce their good behavior with a gift. At least that's how it would look from our perspective. I suggest we review the timeline, as it is now, after this first change. If this change works as intended, we should make it the official plan to take Solaya's timeline and use it as the guide for this planet."

Larissa stepped up to address the Assembly. "Mr. Urthorn, LaDon and I have also been thinking about something else. Between edits, once we make a change and come back to Solaya to see the effects, Earth's history will record us as missing from the point in which we left. If we play highly visible, vital roles, our disappearance might be detrimental to our success. Also, if we continue popping up every one hundred years or so, someone might recognize us, won't they?"

"Jendall and I have been thinking this as well. What do you suggest?" Phelix narrowed his brow as if hoping she already had the answer.

"We thought why not die as martyrs. Stage our death, at least for this first edit. Then, for the next edits, find a way to be no-name scientists who just come up with brilliant discoveries and let the heads of their nations take credit. For the final edit, we put ourselves in power so we will have a platform in order to tell them who we are."

"This makes sense, Larissa. This way we can go in and out of their timeline unnoticed. These edits will need to stretch out over the next three hundred years of Earth's timeline. We should have thought of this before, and we wouldn't have to go through the trouble of staging your deaths. Maybe Jendall and Phelix, since they weren't in positions of noticeable power, can simply disappear without a trace. All families are back on Solaya and accounted for, yes?" Barton asked.

"Yes sir. Everyone is here, safe and sound," Blaine confirmed.

"Very well. All right, let's get this death staging over with. We don't want to start gathering timeline data until this mission is complete. Hopefully we can use your deaths as a way to bind the relationships even further. Especially if the Collectors are deployed correctly to absorb the nuclear blast that might still occur," Barton said. "We want this to be believable. Make it good."

Staging their deaths didn't turn out to be as hard as they had expected. A few pictures, a few reports of missing bodies, and an eye witness stating they spotted someone leaving the building carrying weapons. Not only did this satisfy the masses that two highly respected figures had vanished, it also unified those looking for peace. Just as the Forgotten Wars had proved, those who desired peace started to outweigh those that were against it. Those desiring peace and prosperity began to act out against the protesters. Some protesters were even executed on the spot.

The supposed deaths of Larissa and LaDon's personas quickly became known throughout all the separate factions on Earth. Their deaths became a symbol of unification, not to mention the discovery of such a wonderful invention as the Collectors. They knew this would be the proper defense against those who wished to wage war and profit from their victories. A nuclear attack could now be easily thwarted with this new invention, rendering the bombs practically useless. The team now turned their focus on the outcome of their deaths along with the invention of the Collectors.

Plielix addressed the room as they awaited the arrival of the compilation team. "Jendall and I have just finished gathering the data from their networks. We went about one hundred years into the future. This is well after the first recorded nuclear launch. The launch still occurred, but we will get to that later. There is only speculation at this point as to the structure of the nation until the other team finishes compiling the information now. From the looks of what we gathered, out first edit worked. The launch may have occurred, but our edit deflated the use of nuclear weapons, or at least thwarted them from seeing any effects of launching them.

"They have a fraction of the information to compile than they did originally, so it shouldn't take them near as long as before. Once we see the data in a time based format, we can watch how our initial change unravels within their nation and go a little more in depth concerning the effectiveness of the Collectors."

"Agreed," LaDon said. "And if it lines up appropriately, then we know our plan will work. Each event we impose on their timeline should ultimately produce the same outcome as our own history. All of the pieces are there. Their races have been divided. They are at war just as we used to be. Even the number of separate factions are the same. Five Solayan nations and five Earth nations. It's almost uncanny, even unbelievable in fact. That's a recurring theme that keeps coming up when I study these Earth people. It's almost like looking into a mirror or some type of divination device, watching Solaya's history unfold, but with different characters." LaDon snorted in disbelief at such a plan creating a finely tailored outcome. "It's almost as if...no, never mind."

"No, what, LaDon?" Larissa asked curiously. "You and I have discussed this at length. It's not a crazy idea, LaDon. Say it."

LaDon noticed the Assembly looking at him with great interest, curious about his response. It's as if they were looking for an answer to this exact question.

"Well, it's just that things are going too

smoothly. These are living beings, capable of making all kinds of decisions. Even from the very beginning of this project, before we even found Earth, things just seem to be so ironed out. It's as if there's an extraordinary force at work. I know this sounds like the ideas of some kind of conspiracy theorist. I hope you all agree, that is definitely not what I am. But look at the facts. Finding Earth, finding out that their world is in danger, having the technology and resources available to save them, and the knowledge of our own history superimposed onto their timeline to make it all better. It all seems so... ...so..." LaDon sighed and searched for the right words. "How do the Earth people put it? God-like? Almost as if some supernatural force is guiding all of this into focus, like a bomb seeking its target. I've always believed that fate is an illusion, but now, somehow, I'm not so sure. If anything, this project has shaken the very foundation of my way of thinking."

During LaDon's speech, Barton made his way down to the Observadome floor to address LaDon's question.

"I agree this entire project has lined up perfectly," Barton said. "All the right pieces in place, all at exactly the right time. What you must realize is this project has been the focus of the entire planet. Every resource available has been assigned to this venture. It's our only hope of survival. Just as the memory of the Forgotten Wars has so recently reminded us, when Solayans put their heads together, they will find a solution, no matter the odds. The benefits of such a structured, harmonious nation is exactly what we have. This is the same harmony we are trying to give the people of Earth. Living beings want peace. Deep down they thrive on it. But some get side tracked looking for this peace. They think in order to obtain peace, they must first endure some type of hardship. So I say, let's give it to them. Chaos and bloodshed during the Forgotten Wars were needed to bring the peace that followed. Eventually, people get tired of seeing the blood of their people scattered across the planet. Either that or they destroy themselves. The evidence is already showing itself on Earth. Those who desire peace are actually executing those that oppose them. This is exactly the road our Forgotten Wars were built upon. Our nation was literally built on the bones of those who resisted peace. Solayans were tired of fighting. It was a nasty time in our planet's history. Unfortunately, if our plan starts to unfold properly on Earth, you just might witness this. But understand it is necessary for the betterment of the people. Not everything is beautiful. Compared to Earth, Solaya may be a peaceful nation, full of technological wonders, but it came at a price."

Just then, the doors opened and the compilation team entered with the data. Ramil gave LaDon a sly grin before he exited with his team.

I wonder what that meant. Did it work?

Barton motioned to LaDon to come and retrieve the data brought in by the other team.

"Did it work, Barton?" LaDon whispered as he leaned in to take the device.

Barton gave a look of satisfaction only visible

to LaDon. He didn't say a word, but LaDon knew the outcome. As LaDon made his way to his terminal to display the information for his team to see, he once again was amazed by the skillful weaving of their plan. His lines between coincidence and fate blurred a bit more, leaving him on the fence. LaDon loaded the information into his terminal, which revealed the new timeline. The team took time to analyze the data along with the report from the compilation team. They studied everything in depth, from the introduction of the Collectors to the disappearance of LaDon and Larissa from the timeline. It seemed speculation took over concerning their disappearance. Some said they ran off together. Others said they were presumed dead, although no trace of them was found. From their deaths arose a strong alliance between the Americans and Europeans. So strong, in fact, that they were able to hold back the resistance which was keeping their two cultures apart. Some that were defiant were even converted while others left to join other factions.

The Collectors did their job as well. Although the Australian culture, which had infiltrated much of the North West region of the Americas by the year 2130 A.D., decided to launch the nuclear attack anyway, the Collectors were built and ready by then. The Collectors absorbed most of the blast as well as virtually all of the radiation. With their combined strength, the American and European nations obliterated the Australian population, leaving only four surviving nations across the globe. Aggressive behavior would not be tolerated in a world aiming for peace. With the heads of both the Americans and the Europeans already preaching world peace, they made sure all other factions knew this technology was due to their alliance as well as those who were willing to die for such a cause. In a show of good faith, the Americans and Europeans made the technology public, rendering all nuclear weapons from that point forward useless. However, from that point in the timeline, the team realized that peace only hung around for a fixed amount of time. Eventually everything drifted back into chaos and wars ensued.

"Obviously, everything went as planned. It's time to implement the main part of this mission. It's all or nothing from this point. Here is the timeline I have devised showing the specific events in our history which we will mirror onto this nation. It *will* work. It's going to work," LaDon said with conviction. "As you can see on the timeline, the next discovery to come into play was matter displacement. This science taught Solaya to move matter from one part of the world to another. Once we learned how to break matter into energy long enough to transport it and put it back together on the other side, this taught us that we can assemble matter based off of the original. This brought about matter replication."

"Right. First we show them how to transport, then we show them how to create matter from the elements. This will cause the economy to shift due to the abolition of world hunger. That's one thing I could not get over when we were there those first five years of this journey. A world where there are individuals without something to eat. People dying simply because they didn't have food or water." Phelix looked down in anguish.

"Well we are going to save them, my friend." Jendall reminded him. "And hopefully save us in the process."

"We will, my brilliant scientist. I promise you. If it's the last thing we do. We will save as many of them as we can." LaDon's voice sang a different tune, which caught their attention. "Our timeline has brought us to this point. We can't try some crazy attempt to edit our own timeline. To do so would mean the end of everything we've done. If we were to somehow change our own past, we might not make it to the point we are now. This is a point of no return. We've altered this planet's timeline and we must complete the job. Not only are these poor souls counting on us, our own people are counting on us. Here are the edits. We do them in this order. Each edit has an objective. We must get each of these sciences into the right hands, to the right factions, at the right times. We've proved that we can do it. Let's do it again."

One by one, the team skillfully executed each edit. They checked behind them to make sure the timeline was molding correctly. The matter relocation changed the way Earthlings traveled, minimizing the fuel and resources required. Next came matter replication. Food, water, and other resources could now be created if given the proper elements. This affected the global economy as well. As these two sciences in particular begin to flourish, the need for money started to slowly disappear from existence. People began to see the world differently. Instead of fighting for what they needed, they simply chose their own path. This gave birth to a new age of thinking.

After these two discoveries came many more. As the team completed each edit, they saw Earth following the same path as Solaya. Even the government started to shape itself exactly like Solaya's government, formed many years ago. As each faction grew closer in spirit and harmony, more and more of the planet began to see the rewards. Each culture staying proud of its traditions and heritage, but still open to supporting each other and the traditions of their neighboring allies. Discovery after discovery, one by one, the team found themselves relating more to the Earthlings. Even to the last few drops of bloodshed, the team could see these were a people worth saving. Smart, intelligent beings that all wished for the same thing. Just as Solayans, the people of Earth talked and laughed. They complained about working late hours. They even worried about their own future.

Just as Solaya's history revealed, and right on schedule according to LaDon, they watched the timeline change. These unions started breeding more and more advancements in Earth's nations. The people of Earth began to become competitive. So competitive that they began racing one another. Trying to outdo each other by advancing science more and more. Each race grew more and more proud of their contributions. Demanding respect and recognition, each culture tried to reach higher than the next. This caused small disputes to break out. Eventually, these disputes leaned toward frustration. As a result, these frustrations started to cause a divide, once more, just as it had on Solaya.

"LaDon, why is this happening? Why are they dividing again? All that work, for nothing?" Jendall regarded the last bit of data gathered with disgust. "Can we stop this? I don't understand."

Jendall looked at LaDon with a hopeless expression. LaDon could see the rest of his team looking on with the same expression. The feeling of hopelessness. The feeling that their life's work was being undone by these ungrateful, socially biased people.

"I thought they were better than this? Is it simply in their nature to breed such violence?" Phelix asked as he slammed his fist on the table and rubbed his other hand over his face.

They all looked at LaDon, whose expression hadn't reacted to their pleas for hope. Larissa walked over to his side and stood in front of him, making him look at her.

"Why won't you say anything? What's wrong?" Larissa asked, growing more and more concerned as she watched the look on LaDon's face change slowly. "What? Talk, LaDon!"

"Us." LaDon looked toward the Assembly. "We are the answer. Us four, right here, right now."

"What do you mean, us? Are you saying we can fix this? We can't go back and edit anything we've done up to this point. Where would we start? It wouldn't make sense." Larissa said with desperation in her voice.

By now, all the Assembly members stood at their table looking down at him. He could see the answer standing right in front of him. In their faces he saw the answer.

"Larissa, turn around." LaDon beckoned to her to turn and face the Assembly.

"Okay, I'm looking at the Assembly," Larissa said with a snip in her voice.

LaDon ignored her stubborn attitude and addressed the Assembly from over her shoulder.

"You all were there. When Solaya started to dip back into chaos, battling over technological advancements, you were the Lead Representatives. You were not known as the Assembly at that point, but you were here. If there is one story I know well, it was the formation of the mighty Assembly. My grandfather, Pomph, told me many stories, but this one I requested most of all. Pomph always repeated the fan favorites if I was persistent enough." LaDon felt Larissa's posture change, moving from aggravation to intrigue. "You know how this ends."

LaDon, Larissa, Phelix, and Jendall watched as the Assembly members made their way to the Observadome floor. Jendall and Phelix walked over to join LaDon and Larissa. Both Jendall and Phelix looked to LaDon with expression that told LaDon that they were all playing from the same sheet of music.

Before the Assembly arrived to the ground floor, LaDon faced his team to explain. "It's us. Us four, I mean. We have to be Earth's Assembly. We have to be the governing body for these Earthlings. We have to control the battles, these struggles for intellectual superiority, just as our Assembly controls ours today. It all makes sense now. The teams, the separate groups, all of them working on separate parts of the project. Also the diversity of each team. No one team is made up of one race. It's always a mix, never allowing one race to receive the credit. They keep the balance. I've always known this, but seeing it from this perspective makes all the pieces of the puzzle come together."

LaDon realized through most of his explanation, the Assembly was already gathered behind him. LaDon turned to face them, with what would seem like a sigh of relief.

"All that talk a few years ago about thinking ahead. Let me guess, you're already on this page?" LaDon said with an exhausted look of humor on his face.

"We haven't been on this page as long as you might think, but yes, LaDon. We knew. We see in each of you the same abilities as ourselves. You know this planet. You know its people. You've lived among them," Barton explained. "But it's not over yet. You must go to Earth. Weave yourselves into their nations once more. You must gain their trust. I was able to achieve this trust through patience. Patience with the people of Solaya as well patience of the other Assembly members you see before you."

As Barton said this, Blaine and Alex adjusted their stances, nodding in affirmation. Aleen gave

Barton a small slap on the shoulder as she chuckled to herself.

"We became a family at that point. We trusted each other. To your advantage, I believe you already have that trust with your team here. You must bring the people of Earth even closer than they've already become, LaDon. Make them see how they are slipping, once again, into a history that does not bear fruit. You will always have us here on Solaya to guide you if you ever need advice, although I am sure you can handle it. You say you remember the stories surrounding the formation of the Assembly better than anything?" Barton ended his thought with a question as he had done so many times before.

"Yes sir. Every last one of them," LaDon replied.

"Then you know how it was done. Simply recreate those stories, only in this situation, there are four nations instead of five like here on Solaya. It shouldn't take all that long. You have already laid the groundwork. They should see the error of their ways as Solaya did so many years ago. They should even allow you to control the sciences at that point, as long as you make them realize these advancements are tearing them apart and must be monitored," Barton said with the same familiar confidence that LaDon had grown to love.

"You mean we get to be the Assembly on Earth?" Jendall asked. "I get to be Barton!"

"Hah! No, I believe we all know who fits that role better than anyone. Besides, Jendall, you could never measure up to my wit," Barton said through the laughter as he clasped his hands over Jendall's shoulder and shook him a bit for good measure.

Chapter 23

New Year's Day

If there was anything LaDon knew best, it was the stories surrounding the creation of the Assembly. He could not believe he was going to get to relive those moments, not through a Solasphere, not even through his grandfather's stories, but through his own eyes. And of all the things he had been a part of throughout the entire project, this would be the simplest task of them all.

As a child, he'd acted out these scenes in front of his grandparents. They would laugh until they cried as he would shift from role to role, mimicking each Assembly member. He would start by acting out Blaine Steele's attempt to get everyone's attention before a meeting, showing the frustrated looks on his face and the heavy breathing as he tried to maintain his temper. Then he would switch to Alex Cuberly's piercing glare whenever someone would approach the Assembly with some absurd, mundane request. His passionate side would come to the surface as he wore Aleen Fabian's worried looks on his face as a random delegate would toss out an accusation or make an uncomfortable remark. Of course his personal favorite person to imitate was Barton Urthorn. He remembered pacing back and forth in front of Pomph, quoting some type of logical truth, while slowing his speech for dramatic effect. With all of this in mind, LaDon knew he would still have to be himself and only use these memories as a guide in order to accomplish the task at hand.

"All right, Phelix and Jendall, are you with me? Larissa?" LaDon looked to his team with hope and confidence.

Each of them awaited LaDon's instructions as they had no idea what awaited them, but could see that LaDon knew exactly what he was doing.

"We are going to want to plant ourselves about ten years prior to these feuds. This will give us time to rise into their inner circles by advancing technologies a little more. I suggest one of the tools we use is the second use of the Collector. Now that the technology is well integrated into their way of life, let's show them another way of using it. Showing them that the energy can be collected and then redistributed in a controlled manner will allow them the same benefits we have. The cost of space travel should decrease dramatically as well as the cost of travel in general. Space exploration will take a huge leap. A few years later, with fuel costs lowered, we will introduce the hover technology behind the Solaspheres, and then the Solaspheres themselves. This should give each of us a way to make ourselves known within our nations. Along the way, we must take the credit of each invention together. This way we show them that these inter-societal struggles are not the answer, and they are tearing at the very fabric of their existence they fought so hard to achieve. We can't let them unravel everything they've accomplished up to this point; everything we've accomplished, for that matter." LaDon paused to tend to the puzzled faces surrounding him.

"So, we go back in, put ourselves into positions of authority as we did in the beginning, and then begin introducing more technology? All the while, we encourage them to stop competing with one another for the credit?" Jendall attempted to sound out the plan.

"Exactly," LaDon said, hoping it sank into Jendall's head.

Larissa asked, "But how will this help us create an Assembly?"

"I was almost to that point, but I stopped because you all were looking at me as if I were crazy. As I've said before, if there's anything I know more about history, it's the formation of the Assembly. Technology was much more advanced by then. Solaspheres were just starting to be utilized. History was also being well documented. This is actually my favorite point in our history. It's what made us who we are today. the Assembly brought us together. They taught us that although there may be a small spark of violence within each of us, we cannot deny the accomplishments we can make when we pull together. We don't have to love each other, but we can respect each other enough to get the job done. The people of Solaya back in those days took this to heart. the Assembly has kept it this way. They do not act as dictators. That is not their role. They are governed as well. The Lead Representatives do their jobs, too. This structure within our nation keeps these struggles at bay. Sure there's an occasional

Joh Lin in the bunch, but life wouldn't be as interesting if there weren't."

LaDon felt inside that he was preaching, but seeing that his team was intent on his words, he continued, "Maybe I failed at explaining it. Let me try again, but this time I will make it to the end. We will go in ten years before the feuding begins. This will give us long enough to get ourselves into a position of power. Then, as new sciences develop, we start to introduce some science of our own. While we are introducing our discoveries, we introduce them together, as a unit, in an attempt to show the people that this struggle they are creating doesn't have to be. This will not be hard to accomplish as we have done it before. By then we will be the Lead Representatives due to our discoveries as well as a deep understanding of their past. We created it, so who better to explain it to them? Finally we can introduce the idea of an Assembly. An Assembly made up of us four, but heavily focused on governance by the Lead Representatives. To them this will be a new body of international government, but this new body will ensure the correct individuals get credit while acting as a buffer between the factions."

With the team completely on board, LaDon knew exactly where to start. He found the perfect spot, ten years prior to any feuding. For ten years the team worked diligently, moving up in the ranks, and gaining the respect of the people of Earth. Already understanding their way of life, LaDon and his team were able to maneuver their way in and out of obstacles, knowing that at any point, should something require a change, they had the luxury of coming back to Solaya to change any mistakes that hindered the development of this new nation. As the struggle for scientific superiority began to emerge, LaDon faced the challenge head on, taking advantage of each moment, trying to make the people see reason. Day by day the struggles made themselves more and more prevalent, and day by day, LaDon used these struggles as a platform to promote peace. The Americans started to listen to him. Larissa, Phelix, and Jendall all applied the same logic, standing next to LaDon at news conference after news conference. The ten year fight took its toll on the team. Finally, the people of Earth started to listen to reason. Each faction of Earth, once again taking their cue from their own history, began to rise up against those willing to continue these interracial struggles. This gave LaDon the sign he had been waiting for. Each team member began talks of forming an Assembly to unify the races in order to take this peace one step further. LaDon knew this would solidify the moment and make it stick. With the governments of each culture showing unanimous favor for the idea, LaDon called the team together for a private meeting.

"It is time. We tell them today. We show them that without a governing body such as an Assembly, there is no controlling these feuds. They trust us. We show them that it takes us coming together to overcome these struggles. To unite as a people, still staying proud of our own cultures, but looking upon these discoveries as a moment to celebrate. Not to demean others or put anyone on a higher pedestal just for discovering something first," LaDon said firmly as he faced the team.

"I agree. It was a moment in history to witness the European people coming together. They are all ready for this," Larissa answered immediately, standing from her seat.

"I'm on board. I know the Asian culture; although probably the most proud of their heritage, they are done fighting. They want things back the way they were just after the wars and before all this nonsense," Phelix said as he stood as well.

"Let's make this happen. The Africans are a peaceful but proud group. Like the Asian culture, they just want this to be over," Jendall said, springing to his feet.

The team exited the facility with one task in mind. They each pulled together their governments to a central, discreet location and began the process of forming the Assembly. The debates were few but the process proved to be long. Days went by as decisions were made and laws were passed. Questions came up, such as how to govern the Assembly and what were the roles of the new Lead Representatives. With LaDon and his team now taking their places on the new Assembly, talks ensued about who should take their places as Lead Representatives. As agreements were set, each government found themselves satisfied with the terms and made the first of many public announcements to the world.

From that day, many years passed as new sciences were born. LaDon and his team took certain members of nations that normally would not have met and asked them to work together because they saw the potential of a new science. All the while, Jendall and Phelix dropped subtle hints to lead the scientists to the right answers. In this flurry of discoveries, and the fifteen years of living with the people of Earth as their Assembly, LaDon felt the people of Earth coming into their own. The world was getting used to having an Assembly. He even began to see small children watching in awe as they walked toward the meeting halls. Children waving, hoping to be seen by the great and mighty Assembly. Each time this occurred, LaDon was reminded of the admiration he felt every time he saw Barton Urthorn. Even now, in LaDon's eyes, the man had no equal.

One child stepped from the crowd, looked toward his mother and said, "That's him, mom! I want to be just like him."

LaDon couldn't help but stop and smile at the young child. Vibrant, young, and passionate about the things around him.

LaDon looked his way, and through the guards he said to the boy, "Son, you keep your spirit alive. Don't let anyone hold you back. One day, you'll be walking into this building just as I am today."

The child beamed with delight as his mother thanked LaDon as motherly eyes tend to do.

Realizing it was late, but thinking this was probably a good a time as any, late one evening, LaDon contacted each team member and asked them to visit him and Larissa at their home.

"LaDon, it's after six. Are you sure you want to do this now?" Larissa asked with yearning in her tone.

"Sorry, Larissa. I won't keep them here long. I know they have families to get back to, but I feel strongly about this and I need them here. I hope they will understand," LaDon said as he heard the first knock at his door.

"I'm sure they will, sweetheart. They trust you. They believe in you just as I do." Larissa kissed him softly as she made her way to the door.

First, she noticed Jendall. Phelix was the second to show his face bringing up the rear. Larissa fixed their favorite drinks, which put smiles on both their faces.

"Friends. Obviously you are wondering why on Earth I have asked you to come so late. We have come a long way in these last fifteen years. This entire journey has brought me to love each one of you. Not to get too emotional here, but I would die for each one of you. But as you all know, we have another planet we must save. Our own planet, Solaya." LaDon's soft tone matched the ambiance of the room.

The team smiled as they realized the message LaDon was about to deliver. Each of them looked around the room at one another, smiling with delight. It was just after Christmas time on Earth. The team had been spending quality time with family and friends, and even new friends they had acquired while living on Earth. They knew this time would come. Each of them were ready for one last charade.

"I suggest we tell them on New Year's Day. We tell them who we are and where we come from. This will give them a fresh start. A new beginning. Their world is going to change, again. I believe with our guidance, they will be ready for it. It will be nice to stop lying to them." LaDon said casually as he awaited a response from the team.

"New Year's Day, huh? It may shake them up a bit, don't you think?" Phelix gently swirled his glass of eggnog.

"It might. But as we've said before, they trust us. They have no reason not to. I say we introduce the new science, show them how it works, and then simply tell them who we are," LaDon suggested.

"I'll schedule a meeting with the Lead Representatives on Monday," Jendall said as he made a note in his viewer.

"Well. That settles it then. I am going to return to Solaya tonight to inform the Assembly of our plan." LaDon's body language said the meeting was closed. "Feel free to stay here as long as you'd like. You are welcome to anything, as you know."

"I believe I am going to head home. The family is waiting. My wife was kind of surprised at this late night meeting call. I told her I figured what it was about. I was right. I'm sure Jendall's wife felt the same," Phelix said as he moved toward the door.

"Yep. That's the story at my house as well. Take care, LaDon and Larissa. I'll make the necessary plans in the morning." Jendall followed Phelix out the door and into their vehicles.

With the night coming to a close, LaDon and Larissa retired to the living room. Larissa sat as close to LaDon as possible.

"Do you want me to go with you to Solaya tonight?" Larissa asked sweetly as she scooted even closer to LaDon.

LaDon pulled her close, still making sure he had eye contact, and answered, "No. I think would like to do this alone. I want to inform Barton about our plans, but I also want to talk with him about the last few years here on Earth. I want to swap stories, if you will. He means a lot to me, Larissa. Sometimes, when I look at him, I can't help but think of my grandfather. Ever since this project, he has looked after me, watched over me. I have come to respect him a great deal more, now, after what we've all been through."

"I agree. I realize each of them has already been through this, and it does warrant respect for them as well as their position." Larissa looked into his eyes with understanding. "As for going back without me? I won't argue with you this time. I understand. Will you stay here for a few more minutes before you go?"

"Anything for the most beautiful woman on Earth." LaDon leaned forward and kissed her lips. "I do not believe any of this would have been possible without you by my side. You've kept me calm through the storms. You kept me sane through adversities. This planet is lucky to have you, just as I am lucky to have you. I want things to stay just like this for a long time to come. I've loved you for years,

Larissa Sonne, and I'm going to love you for years to come."

LaDon and Larissa held on to their intimate moment until they found themselves making their way up the stairs and into their bedroom.

Closing the door behind him, LaDon sighed and said, "I guess I will go see them in the morning."

Chapter 24

The Truth in Its Infamous Wisdom

The next morning, bright and early, LaDon readied himself to port back to Solaya for a final meeting with Barton and the Assembly. It was night time there when they left, so he realized they would probably be wrapping up their day when he returned. Larissa, squinting to see the time in her viewer, sat up slowly in bed, only to see LaDon fully dressed, getting ready to leave.

"Tell them I said hello and that I've been thinking about them. I kind of wish I was going with you. I miss the smell of the plant life surrounding Nalkalin." Larissa tried to wake herself long enough for his departure. "Here, let me get up and close the blinds."

Larissa started to get up and LaDon sat quickly on the bed to stop her.

"Don't get up. I'll close the blinds. You sit back. When I disappear, I will be coming right back, so don't worry," LaDon said as he felt her relax back into the bed. "I'll only be gone a second or two."

"Yeah, but I know you will miss me much longer than I will miss you." Larissa kissed him passionately, reminding him of last night. "There.

That should hold you until you get back."

Without saying a word, LaDon stood from the bed, closed the blinds, and took two steps back. He pulled his return module from his drawer and with a flash, his vision blurred and his surroundings were suddenly changed to the dimly lit Observadome.

Making momentary visits throughout his tenure on Earth kept his memory of love and home, but this visit felt different. Everything felt a bit smaller, even a bit foreign in some aspects. Deep down he felt like this would be the last time he was on Solaya for quite a while. He knew this wasn't true, but he could feel Earth calling him back. He stepped down from his pad, checked the time sync with his viewer, and realized the Assembly was probably finishing up the day back in the Unification Chamber. LaDon strode down the familiar corridors and down the huge hallway. The big double doors, which were once a source of intimidation, didn't seem quite as heavy this time as he opened it slowly, peering into the room. He saw only one figure in the mostly dark room.

"Excuse me?" LaDon said, trying not to startle the figure. "It's LaDon."

LaDon recognized Barton's voice immediately as he answered, "LaDon! What a pleasure! Of course, you only left a few moments ago. Ten years has done you well, I must say. Where's the rest of your team?"

"It's just me, sir. Don't worry. I've already created new return modules for them so they don't return before I do," LaDon quickly explained. "I just came back because I wanted to talk to you."

"Come. Sit." Barton motioned for LaDon to take his personal seat at the meeting table.

"Oh, sir. I couldn't."

"Come now. You've earned it. I can see the years on your face. You've earned this seat," Barton insisted. He took Aleen's seat and LaDon took Barton's.

"Well, sir. After all we've been through, we made it. We have formed the Assembly on Earth. Everything worked as we'd hoped. We have decided to bring forth the technology of interspatial travel. Once we prove its functionality, we will tell them who we are and explain why we are there." LaDon watched Barton's reaction. "This will mean much for Solaya, so I wanted you to know so you can properly inform Solaya of the news."

LaDon watched intently as Barton leaned forward in his chair, all the while looking at LaDon with a sense of peace behind his eyes.

"I am proud of you, LaDon. We are proud of all of you. You have a lot still ahead of you, as do I and the rest of the Assembly here on Solaya. With that in mind, I have something I wish to share with you." Barton reached into his pocket.

Barton pulled out a storage device similar to the same kind they used to gather reconnaissance on Earth. LaDon took it and looked at Barton with a puzzled face.

"This storage unit is classified. It is for your eyes only," Barton said as he moved his eyes toward the device.

LaDon looked down to see the device was

labeled. The label simply read 'LaDon.'

"I want you to take this to your office, use that glorious desk we built you, and view the contents. This storage unit has been adjusted to integrate into our terminals. All of the files are time stamped and labeled. Some files are images while others are Solasphere recordings. Watch them in sequence and *only* in sequence; otherwise, they won't make sense."

"What's on it?" LaDon asked reflexively.

Barton stood from his seat, ushering LaDon to the door.

"Go. You should find that out for yourself." Barton led LaDon out the door. "You're a good man, LaDon Grafter. You come from a great family. I love you like a son, and I always will."

Barton's words rang in LaDon's ears as he slowly made his way to his office. Unsure of what was on the drive, right now, he could only speculate. By the time he reached his office, he was done speculating and simply wanted to see what was on the drive. He activated his desk, which came to life in brilliant color. LaDon was reminded how he missed the latest and great technology while living on Earth, but that thought was quickly snuffed out as the contents of the drive showed up on the interface.

LaDon picked the first file, which was an image file, and opened it. The photo seemed to be rendered in older technology, as the crispness of the image was a little distorted when compared to today's technology. He looked at the time stamp. This proved his assumption.

Solaya, Assembly Meeting, Year, 3830? That's

nearly three hundred years ago.

The photo was of a man of spectacular build. He was in shape for someone whose aged face in his fifties told a different story. He was surrounded by a woman and three other men, who all seemed to be sitting at a table, much like the Unification Chamber. The next file was a recording. LaDon noticed the time stamp was the same as the picture. It had the same time with a similar title. There was no physical interface capabilities with this file; therefore, LaDon immediately realized this file was obviously older since he could not smell or feel anything. He could only watch and listen. The recording showed this same group of individuals. This time they were debating something. LaDon ran it back to listen closer as his mind was distracted at the beginning.

"We cannot continue fighting amongst ourselves. We are getting nowhere. Can't you realize, the small alliances we have been able to form have delivered us from the dark ages. Technology is taking leaps and it's all because of these unions. We have come too far to quit now." The recording came to an end.

LaDon's mind raced. He did not quite recognize the faces in the picture and on the recording, although the voices were obvious. *That was Barton's voice, but that man looks nothing like him.* LaDon quickly moved on to the next file. The next file was an image, similar to the first, with a matching recording, both named the same thing. *Solaya, Assembly Meeting, Year, 3980.* LaDon opened the next image. This image was similar to the first. A group of five individuals, each appearing to be debating something with an unknown group sitting in front of them. LaDon opened the matching recording and listened.

"*Representative, please. You must understand, we are not trying to dictate. We are trying to keep the peace. You are our check and balance, you know this.*" One of the people from the photo spoke passionately to the crowd in front of them.

The recording stopped. *But this is nearly one hundred and fifty years ago. Those don't look like the same people, but they sure sound the same. Also, they may look different, but they are all similar to the first group.* LaDon looked ahead to see each file name and time stamp. The files started to show a pattern. These recordings went right up to the moment the Assembly announced to Solaya that interspatial travel was possible at the big meeting. Against Barton's orders, LaDon skipped ahead to one of the recordings of that meeting. He watched a small snip of the meeting he attended years before. LaDon went back and started with the third recording. He noticed each time he listened to a recording, the voices were those of the Assembly. He was sure of it. *But these recordings were made hundreds of years ago. How could they be in so many different points in time at once? They would need...no wait, that's not possible.*

Like falling from a cliff, a rush adrenaline filled LaDon's body. If his heart were a bomb, he would have been dead. *It can't be,* he thought to himself. LaDon knew there was only one place he could see this information properly. He sprang from his desk without even deactivating it and ran as fast as his legs would take him. He rushed up to the lift, frantically pressing the button labeled OD. Once the doors opened, LaDon pushed his way out and lunged at the door of the Observadome. He cleared the small flight of steps in one leap, landing on the Observadome floor with a thud, and raced to his terminal. As fast as his fingers would go, he configured the terminal to display each file, setting the recordings to loop. Each screen displayed a file based on its timeline, side by side, around the room, in a three-hundred-and-sixty-degree view. LaDon stood in the middle of the room, turning in circles. Slowly turning, he could see each face in the images and the recordings. They were all similar but different. He watched as the faces slowly turned into the faces he was most familiar. Barton, Aleen, Blaine, Alex, and one more unknown member of the Assembly only found toward the beginning of the data, although they did not appear in the later ones. *It's true. They've been moving through Solaya's timeline, editing our history.*

LaDon asked aloud without thinking, "But why?"

"To save you, of course." Barton's voice spoke out from the darkness as he slowly walked toward LaDon, who looked to Barton in astonishment and disbelief. "To save this planet and to save Earth in the process. Both of your planets were in trouble. The people of Solaya needed a place to go. The people of Earth needed a savior. They needed *you*, LaDon."

"Me? But...the timeline...you have to be

outside the timeline to make all of this possible. It would be chaos trying to go back and forth from Earth to Solaya, remembering what to edit, how to edit, I mean..." LaDon tried desperately to apply logic.

"Slow down, LaDon. Sit with me here." Barton dragged two chairs together. "Now. Ask. Ask your questions and I will tell you everything you want to know."

"Who are you?" LaDon asked simply.

"My name is Barton Urthorn."

"Where are you from?" LaDon tried again, in a simple tone of voice attempting to contain his emotions.

"I am from Yarin Four," Barton answered.

"Yarin Four? Not in this universe or Earth's universe, correct?"

"Correct," Barton said plainly.

LaDon looked into Barton's eyes and said, "Heh, I got that one right. I knew you couldn't be from either universe."

"Please, continue. I gave you this storage drive for a reason. I want you to know it all," Barton said, which drew another question from LaDon.

"Why tell me all of this?" LaDon asked with wonderment behind his eyes.

"Because tomorrow we're about to tell Solaya exactly what you are going to tell Earth in a few days." Barton watched LaDon's face closely.

"You mean, the Assembly is about to announce to Solaya that they are from another planet?"

Barton acknowledged his question with a firm nod.

"I have another question." Barton motioned for him to proceed. "Why me? Why not someone else? You said they needed me."

"You were our finest discovery, LaDon. In your viewing of the data, did you notice anyone that didn't quite fit the picture you were trying to portray in your mind?" Barton asked.

"Yes. The fifth man. I recognized Blaine and Alex. Of course I recognized you and Aleen, but there was a fifth person that did not seem familiar and neither did his voice," LaDon said as the puzzle remained printed on his face.

"That man was your father, LaDon. Haven't you ever wondered why there's only four of us? There's supposed to be one from each nation, correct?" Barton asked LaDon as he leaned down to meet his gaze.

"Well, yes, but it never came to mind to question it. I just figured you hadn't picked anyone out yet." LaDon answered honestly as his mind began to turn once more.

"He held the Vaknoreeyan position. We needed to replace him. We asked your grandfather, Pomph, if he would join us. He simply said no. He didn't enjoy the political side of things. He was a historian, and he was good at it too. LaDon, when your father passed and your grandparents promised to take care of you, we were still in the early stages of repairing Solaya. We were trying to get everyone ready for this journey you are on right now. We didn't know exactly who would be able to handle the task of repairing Earth. When Pomph recognized your love of history, he came to us, explaining just how vibrant you were and how much you loved stories of the Forgotten Wars. It was then we knew. With your vast knowledge of Solaya's history, who better to mold Earth's timeline than you? As I said, LaDon, you were our greatest discovery. No technological advancement could match your abilities." Barton stopped to catch LaDon's reaction.

LaDon's eyes were glazed over with tears as he looked up at Barton and smiled.

"I had no idea."

"Think if you had known. What might have happened?" Barton checked to see if LaDon was following his logic.

"Oh, I get it. I understand totally." LaDon said as he could feel another question coming on.

Before Barton could say anything, LaDon stepped in with another question.

"Another question then. Can I ask you how my father died? I am beginning to doubt it was from a deep space accident." LaDon looked up at Barton who appeared moved by the question. "Barton? Are you all right?"

"Your mother did, in fact, perish in the deep space accident. Your father, well, we…" Barton turned his face aside to cover his emotions as his breathing began to stagger. "We lost him to a return module accident, LaDon. He picked up an old one. Before any of us noticed it, he was gone. It's so rare that we lose someone to such a thing. He was a great man, LaDon. A great man! And don't you ever let anyone tell you different. It was such a stupid mistake! I blame myself to this day."

Barton stood from his seat and walked a few steps away to collect himself. LaDon gave him his privacy as watched the man he admired most in the world weep for a lost friend.

"I'm sure it wasn't your fault, Barton." LaDon finally spoke as he felt Barton start to calm.

"We're just so careful about that kind of thing!" Barton growled as he shook his hands in frustration as he turned to face LaDon. "I loved him, LaDon. He was my best friend. I vowed that day that I would make sure nothing ever happened to you and you would have anything you desired. When Pomph promised to take care of you, I knew we could continue our work here on Solaya, and I could keep an eye on you from a distance. I had no idea until you got older that you would fill the shoes of your father. You talk about fate, LaDon. That's fate, my son. No matter who's controlling the timeline. It exists and your father is living proof. Fate put the Grafter family in my life. I'll believe that until the day I am dead and gone."

"Sir, may I ask another question?" LaDon said sheepishly.

Through his tears, Barton grinned. "Of course, LaDon. I told you I'd tell you everything."

"You say you're from Yarin Four and you keep saying things like 'we are always careful' and 'when we lose someone'. You say this in a strange context. Can you make that make sense for me?" LaDon asked as Barton made his way back to his seat to face LaDon.

"Yes, we are from Yarin Four. So, you caught that, did you?" Barton said with a humorous huff obviously realizing his slip. "Well, I said I would tell you everything. We come from a planet whose sole purpose, in the beginning stages, was to search for other planets containing life, watch their history and study their ways of life. We hoped to learn more about ourselves to find better ways of living. Maybe some intelligent species that had figured something out that we had not. But from that, we realized many planets did not make it past the technological adolescence as we did. We were lucky. Our history is much like that of Solaya and Earth. As time passed, we found that we could edit the different planets we encountered, save them from their destruction, and add their lives to our own. In the end, we're all alone out here in this vastness, and to be able to link us all together became our goal. We call ourselves the Universal Menders. This is my team's final journey. Solaya is where we live now, and Solaya is where we will die. Earth will be that place for you."

LaDon took a moment to digest the information he had accumulated over the past couple of hours. He imagined what it was going to be like when he told Larissa. Jendall and Phelix would come apart with excitement. They would all want to come back to Solaya to hear all about Yarin Four and talk with each one of the Assembly members. He also took a moment to himself. He looked to Barton, who was sitting quietly, obviously giving LaDon the time he needed to soak in what he had learned.

"Well, I suppose I still have a job to do, and you are probably tired from the day," LaDon said as he looked at Barton who was staring back at him in contemplation.

"Yes. It's been a long day. Let me be the first to wish you luck. Facing those people, telling them who you really are, isn't going to be easy. We are actually in the same boat now, you and I, and boy, do we have a story to tell." Barton laughed, breaking the tension in the room. "Oh, by the way, how are things with you and Larissa?"

"Ah, things are going great! She's the wonder of my world." LaDon looked at Barton. "Now don't tell me you created our relationship with edits or something like that. Tell me it's not true."

"Nope. That one was all you, my dear boy. All you," Barton said with a smile.

Both of them stood as LaDon entered the combination to return him to the point in which he left. He started toward the pad and turned back a moment to face Barton.

"Barton?"

When Barton turned, LaDon closed the gap between them. LaDon looked at him for a moment and an overwhelming urge to cry overtook him. His eyes welled up with tears. It was as if Barton could read the expression on LaDon's face. Barton threw his arms around LaDon and pulled him in. LaDon returned the hug, happy no words were needed. With a tear rolling down his face, LaDon backed away and looked once more into the eyes of the most brilliant man he'd ever known.

"You're a good man, LaDon. It's in your blood, kid." Barton patted LaDon strongly on the shoulder.

LaDon shook his head as he wiped the tears from his eyes. He walked over to the pad and stepped up on the platform. He looked to Barton, who was standing at the terminal LaDon just configured. With a flick of his wrist, Barton flipped the data drive to LaDon containing all the information about the Assembly.

"You almost forgot this. Just don't hit 'em with it all at once, huh?" Barton said in the same wise tone of voice that LaDon had come to cherish.

LaDon watched as Barton activated the sequence. With a rush of emotion and a sudden welling of new tears, his vision blurred. He was not sure if the tears or the temporal distortion caused the blurriness, but in that instant, LaDon watched Barton disappear from view and Larissa suddenly appear in his place. She was still lying in bed, just as beautiful as the moment he left her. At the sight of his tears, Larissa sprang from the bed and rushed to LaDon.

"LaDon, what's wrong? Are you hurt?" Larissa asked frantically.

"I'm fine." Her warm hand wiped away his tears.

"Then what's wrong? Why are you crying?" Larissa asked again with the same frantic tone.

Calmly, LaDon placed his hands on Larissa's shoulders and said, "Larissa, sit down. You are not going to believe what I am about to tell you."

Come visit me on Facebook for news about my next book, or drop by and let me know what you think

https://www.facebook.com/lukeanthonylang

Special Thanks To My Beta Readers

Chris Roberts
(Motivator)

Marin Vucić
(Scientific Advisor)

Jamie Davis St. John
(Relationship Instructor)

David Short
(The Real Truth)

Lauren Howle
(Character Development)

About The Author:

Luke Lang works a forty-hour–a-week job posing as a run-of-the-mill IT guy who gets the luxury of programming from time to time. He has had the pleasure of working this job for the past 17 years and wouldn't trade it for anything. He has an Associate's degree in Computer Science and a Bachelor's degree in the Management of Information Systems. He also maintains his Certified Information Systems Security Professional (CISSP) certification.

Luke lives in Calera, Alabama with his wife Jennifer. They have been married for 13 years and have three dogs, Vinton, Ramsay (Ram Bam), and Dallas.

Luke used to enjoy video games until he discovered the joy of writing, although you can still ask him about his many video game achievements. Don't tell anyone, but he sneaks around and plays from time to time, and he is always willing to share his adventures.